Knowing tomorrow?
How science deals with the future

Knowing tomorrow?

How science deals with the future

Edited by: Patrick van der Duin

With contributions by:
Eleonora Barbieri Masini
Erik den Hartigh
Dap Hartmann
Peter Hayward
Cornelius Hazeu
Sohail Inayatullah
Ela Krawczyk
Graham May
Alan Porter
Patrick van der Duin
Joseph Voros

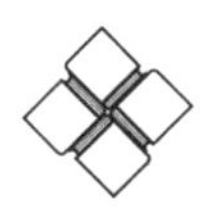

Eburon Academic Publishers
2007

This book is financially supported by the Dutch Scientific Council for Governmental Policy, Delft University of Technology and Alares.

ISBN 978-90-5972-229-3

Eburon Academic Publishers
P/O Box 2867
2601 CW Delft
The Netherlands
Tel.: +31 (0)15 2131484 / Fax: +31 (0)15 2146888 info@eburon.nl / www.eburon.nl

Cover design: Studio Hermkens, Amsterdam
DTP: Textcetera, The Hague

Table of contents

1 Futures research and science: introduction

Patrick van der Duin

1 Introduction

This book is about how in various scientific disciplines the concept of the future is being addressed. Since most of these scientific disciplines existed before the start of *serious* or *modern* futures research, that began just after World War II, it is interesting how scientists of these disciplines were already then addressing the future in a more general sense. For instance, sociologists have showed since the beginning of their discipline, starting with the work of Emile Durkheim, great interest in utopias and utopian thinking in the 17th, 18th and 19th century. And economists started at the beginning of the 20th century to think about how economic cycles from the past were shaping the economic future.

But this book aims not only on giving an interesting historical account on how these disciplines addressed the future. A serious claim is also made to gain more knowledge about the history of modern futures research so that current practices in this field can benefit from this. This might need some explanation because not every futures researcher or someone who is interested in futures research will find it obvious that we can learn from history. After all, history does not repeat itself, as every historian would agree. Rather, by thinking about which role the future plays in scientific disciplines such as psychology, sociology, or astronomy, futures research and its practitioners will become more aware of the *context* in which they are doing their work. That is, studies of the future are not a goal in itself but serve as an input to decisions about issues that are strongly relating to scientific disciplines. For instance, a futures researcher who thinks and writes about the future course of economic developments should definitely take into account the (current) theories, principles, and mechanisms offered by economists. His added value above an economist who would do the same is that the futures researcher is capable of adding insights from other scientific disciplines.

This alsmost automatically makes the futures researcher and subsequently futures research an activity characterized by holism, inter-, multi-, and transdisciplinarity. *Holism* because the inherent uncertain nature of the future means that it is (or will be) more than its constituent parts. Otherwise, the future would just be something as a calculus (which it is not, in my view). *Interdisciplinary* because futures research often focuses on *social systems* (instead of natural systems) that are tend to be open systems in which different societal spheres are connected to each other. This empirical observation should of course be captured by theories that describe these relationships and how they develop into the future. *Multidisciplinary* because societal issues can and ought to be viewed from many different perspectives. Futures researchers should bear this in mind when they are researching the future of the issue at stake. And without working towards a boring compromise, futures researchers should search for a valid, workable, and 'future-proof' integration between these perspectives. And *transdisciplinary* because futures research might ultimately become a discipline that crosses all disciplines (or most) and that combines the knowledge and theories of these disciplines. Therefore, futures researchers should not only possess factual knowledge about a certain domain, as is the case for scientists in most scientific disciplines. But they should focus as well on process skills to make possible to cut through all the classical scienitific disciplines. Futures research then boils down to arranging *Neue Kombinationen*, to use Austrian economist Joseph Schumpeter (see also Chapter 9 on futures research and the management sciences) between different scientific disciplines to develop new insights about the future.

Thus, futures research is (or has become) more possessing a specific competence than having access to a database containing certain factual or theoretical knowledge. But this was not always the case. The first futures researchers were mainly *futurologists* who knew, or claimed to know, much about the future development of a specific subject, often technology related matters. Nowadays these futurologists are still present and wellknown but they are accompanied by futures researchers who focus less on predicting the future but more about how to organize processes in which the future is being *explored*. Future *content* experts, i.e., the futurologists such as Alvin Toffler and John Naisbitt, are operating next to future *process* experts, to use Joel Barkers' categorisation. Both types of futures researchers are, however, not often working together. They are worlds apart. That is, the future process expert might use the opinions and future expectations raised by future content experts but this takes place often on-line, and the use of the expertise of future process experts by future content experts is, as far as I know, largely absent. In a sense, the future content expert is operating from out a specific scientific discipline and often stays within that scientific domain. Economists will probably not make predictions about societal developments. Although some famous futurologists can be considered *generalists* who express their opinions and predictions for whoever asks them and about any topic. Just as economists deal in models, futurologists deal in predictions and future visions. The paradoxical situation arises that futurologists, who apparantly appeal to a large audience and often have a *guru-image*, make futures research *elitist*. The futurologist claims to know about future developments and his very existence means that he (or she) is the only one who knows the future. Remember that the futurologist not only claims this position but it is also given to him by the public and by his clients. Indeed, powerful futurologists can establish their own self-fulfilling prophecies and uphold them. They are the *modern prophets*. To a certain extent, the future process expert and the future content expert are competing who is the best in looking to the future. I think that we could say that the future content expert has won if you take into account criteria such as media exposure, impact on decision-making, and the hour tarrif. However, this victory can be considered a 'Pyrrhus-victory', at least for the future process expert, since their work has given looking to the future a doubtful name. Their numerous wrong predictions, often hilarious and even more often with a big negative impact, are the cause for this.

In a way every expert, that is, every scientist doing some kind of work within a scientific discipline can be considered a futurologist or a futures researcher. Being an expert within a certain field almost automatically means that someone knows a lot about the (possible) future course of his expertise. The added value of a futures process expert above a 'standard' expert is that he or she masters the process skills and futures research methods to do a study of the future with contributions from these experts and by using other data sources. The added value of a future process expert is also that he or she can combine the different fields of expertises into a coherent vision of the future. In addition to this all, I would like to introduce the *future **study** expert* who studies how and why people and organisations are doing futures research. More specifically, they look, for example, at what kinds of methods futures researchers apply, what their motives are for doing futures research, and how studies of the future are used in decision-making processes. This book can be considered the result of the work of future study experts, although some authors do have a lot of knowledge of a certain topic, domain, or scientific discipline. So, this book does not contain future visions or predictions of a certain issue, it also does not describe different approaches to studying the future, and it also does not present a new method of futures research. Furthermore, this book is not about how the future of these scientific disciplines will be *in the future*. This book might be much less instrumental than that. It dives into how different sciences deal and have dealt with the concept of the future. In doing this it provides knowledge about the science of futures research (see below) and an interesting and relevant context in which futures research is taking place. Looking back into history to see how the future was envisaged in those days is not uncommon for futures researchers. Next to looking to the future itself, it might even be the second most beloved activity by futures researchers.

2 The terminology of futures research

Not many scientific disciplines (see the next section) have so much different names as futures research (as I call it; see below). The list of names is long: technology forecasting, technology assessment, futurology, prediction, foresight, longe range planning, futures studies, conjecture, technology intelligence, fuurist thinking, futures studies, technology foresight, prospectivism, prognostics, futurism, futures management, prospective study, and I probably have forgotten a certain amount of other terms. Also, in his review of the current state of futures research in this book, Alan Porter uses many different names including: technology monitoring, technology watch, technology alerts, intelligence and competitive intelligence, technology forecasting, technology roadmapping, technology assessment, and forms of impact assessment, including strategic environmental assessment, and technology foresight, also national and regional foresight.

This long list of different names is not something to be proud of since it is not a proof of maturity and perhaps also not of professionality. I do not think that astronomists often have discussions about what the correct word is for their discipline and I also do not think that there are many different names for their discipline. Nevertheless, the existence of different names used for 'looking to the future' is also a proof that the type of activities attached to looking to the future is changing and thereby reflecting a specific historical periode. For instance, in the 1950s, when society was assumed (rightly, I guess) to be stable and a lot of faith was put in the (rapid) development of science and technology, the term 'technology forecasting' was most used. And right indeed, since then looking to the future most resembled technology forecasting. When the image of technology as benefactor to society started to crumble down more attention was given to the unexpected and often bad consequences of technology to society, the term technology assessment became more in use. However, I have chosen to use the term futures research for three reasons (Van der Duin 2006, p. 39):

- *Multiplicity*: the term *futures* refers to thinking in multiple futures (instead of just one), which nowadays is very common or even dominant in studies of the future.
- *Multidimensionality*: the term *futures* also suggests that possible futures are considered from a social, cultural, economic, political, and technological point of view.
- *Investigation*: the term *research* implies that we do not adopt an *a priori* standpoint with regard to the question whether or not it is possible to predict, create or explore the future, emphasizing instead that the future can be investigated and knowledge about the future can be gained which can serve as a valuable input to today's decisions about the future.

I think that it is a safe prediction to say that the discussion about the right term for 'looking to the future' as well as process of extending the list with new names will continue for some while into the future. After all, a discipline such as futures reseach, containing many 'free thinkers' and in which the future is considered to be a realm of human, societal, political, and scientific freedom, will by nature have difficulties in standardizing its own terminology.

3 Futures research as a science?

Much has been said and written about whether futures research is a scientific discipline or not. Some say that it is not a science since the future does not exists as a tangible reality, only as something which takes place within people's minds. Or, as other's say, the future cannot be predicted so scientific hypotheses about the future cannot be tested. In addition to that, some claim that futures research is unscientific given the huge amount of wrong predictions caused by the inherent unpredictable nature of the future.

I consider futures research a science. Not in the in a classical sense, as characterized by a mechanistic view on the world, a quantitative approach to reality, carried out by detached scientists, that predict the future course of certain variables. But more as a social science, focused on how people deal with future matters, that acknowledges the complex interplay between man, nature, technology, and society. That the future as such does not exist is of lesser importance to me. As a trained macro-economist I know that inflation or economic growth does not exist. They are only concepts formulated by scientists and policy-makers to give some structure to the perceived chaotic reality. But I do not know many people who state that economics is not a science, only those who regard all social sciences as non-scientific. Nevertheless, the simple fact that something as the future only exists in people's minds makes it not less tangible. People decide, act, and react for a great part on what their future expectations and wishes are thereby making the future as real as anything else. Bell and Mau already pointed to this in the 1970s.

Also, the argument of futures research not being capable of formulating testable hypothesis is not correct in my view. Indeed, hypotheses about future developments can not be proven true or false immediately. The future takes some time to develop so to test hypotheses about the future only a sufficient amount of patience is needed. Given the slow time in which science develops (everyone who has ever written ánd published a scientific article knows what I mean), patience is or should be a valuable part of the character of each scientist.

Furthermore, the inherent forward looking nature of futures research might even makes it the most scientific science of all!....The criticism on futures research in being unable to provide good predictions should be accepted. That is also one of the reasons why futures research has made a shift from merely predicting the future to exploring the future. By narrowing down futures research to only predicting the future, the critics of futures research have shown to held an antique view of what this disipline is actually doing.

One could say that the discussion about whether futures research is a science or not is a not very interesting one, Especially futures researchers with much confident (often the future content experts, I think) might be inclined to do so. For them it might be an academic discussion of which the outcome will (probably) not affect their position and their work. Others might say that since this discussion will never reach an agreement this makes the discussion not very fruitful. They might be some truth in there words since it expresses, after all, an individual opinion, almost a matter of taste. If a futures researcher states that the academic status of his work is irrelevant it is wise for him or her not to engage in the discussion about the acclaimed unscientific nature of futures research. If a practitioner of futures research does not expect to gain more assignments as soon as there is consensus about that futures research is indeed a science, it is understandable that the practitioner keeps a low profile. But personally I think that this discussion is important, and maybe my professional background as a scientist plays here a role. I do think that the status and position, in academics ánd business, will improve if futures research is considered a science. And that is something I think all futures researchers, both in academic circles and in business, will welcome. Indeed, gathering knowledge about the (historical) role and function of the future in various scientific areas will enable us to reinforce the scientific foundations of futures research.

In addition to all this, the academic status of futures research will improve as well thereby (hopefully) gaining more resources (money, people, time) to do scientific studies on the future. Also, this might give futures research a place in the curriculum of universities and other forms of higher education. Currently, worldwide only three Master courses in Foresight are offered: Turku (Finland), Australia, and in Houston (U.S.A.). Although I do not think many futures researchers consider it as a pity or a shortcoming of their education that they were 'raised' in an other discipline, I do think that such studies are of great importance. Personally, I would even welcome the idea to give thinking about the future a fixed place in curricula of lower forms of education. Why do children of, say, five years 'know' what they would like to be when they have grown up, and why is it that children of, say, fifteen years have difficulties to decide on what courses on school to follow because they haven't got a clue about what profession to take up in the future? Why do we teach our children history, and why not futures research?

4 Overview of the book

As said above thus book deals with the question how in different scientific disciplines the fure is being addressed. Each chapter deals with a specific discipline. It is of course impossible to cover all scientific disciplines so a choice has to be made which discipline to include and which not. I have decided to put more emphasis on social sciences because I regard futures research as a social science and because the emphasis of studies of the future has shifted from purely technological issues to more human and social matters. Therefore I have included: economics, sociology, politics, philosophy, and psychology. To make clear that the future even plays a role in a 'juxtaposed' science, history is also a topic in this book. The increasing importance of concepts as 'sustainability' and other 'green matters' have made me to decide to include the topic of nature and environment. Although the emphasis is on social sciences, a classical natural science such as astronomy definitely belongs in this book.

Planning has always been an important forward looking activity and that is why geography is addressed here. Finally, a modern and applied science such as management science also deals heavily with future matters.

In Chapter 2, dr. Peter Hayward of Swinburne University of Technology, (Australia) discusses how the future is dealt with in the science of psychology. He investigates how individuals "perceive the future". Hayward takes an evolutionary approach to development of the brain and its 'forward-looking'-functions and focuses on 'future thought'.
The role of the future in sociology is the topic of Chapter 3, written by prof. Eleonora Barbieri Masini (Italy). She adopts a human and social perspective to futures research. She regards futures research as a social science, which includes the issues of values and choices, thereby making it necessary to include a philosophical and ethical perspective. Given that that looking to the future and building a vision of the future is intimately connected with *change* it follows that value choices are inescapable and thus an important topic of futures research.
The role of the future in the economic science is the topic of Chapter 4. Dr. Cornelius Hazeu, formerly working for the Netherlands' Scientific Council for Government Policy and currently working as an independent consultant, goes into how economist give the future a place in their thinking. Hazeu does not only look at the future in economic theory but also in economic policy since both are intimately linked in the economic science. It shows that economists acknowledge the uncertain nature of the future but that does not make them turn their back on the future. Rather, they try to give the unpredictable future a place in their thinking and reasoning by focusing on 'if-then' reasoning based on pre-defined assumptions and looking to the plausibility and probability of their economic models.
The question what philosophers say about the concept of future is discussed in Chapter 5 by dr. Joseph Voros of Swinburne University of Technology (Australia). He focuses not on philosophy in general, but describes the methodological foundations of futures research which are subsequently based on philosophical views. Voros clearly describes what consequences changing methodological paradigms have for doing futures research.
That even historians care about the future is illustrated in Chapter 6 by prof. Sohail Inayatullah (Tamking University, Taiwan). He focuses on macro-historians, by which he refers to those historians who study the history of social systems to find "laws of social change". Certain laws do not only aim at describing or explaining the historical rise and fall of civilizations, but by their macro-nature become capable of outlining the (possible) future of civilization

of culture. In essence, these theories seem to become a-historical, which makes them, in theory, suitable as 'theories of the future'. However, as Inayatullah asserts, the difference between a futurologist and a macro-historian is that the former focuses on the individual ability (and wish) to shape the future while the second puts emphasis on the structural (macro) circumstances in which individuals create the future.

In Chapter 7 dr. Ela Krawczyk (assistant-Secretary General, World Futures Studies Federation) describes which role the future plays in geography, focusing on spatial and urban planning. It is shown that urban planning by nature touches upon issues to do with the future. As soon as geographers started to think about change, the concept of time and with it the future was introduced. Krawczyk provides a clear historical account of how geographers have looked to the future, showing a shift from a ´blueprint´-future towards a more open future containing different alternatives.

Graham May (United Kingdom, Future Skills) discusses the future in relation to natural phenomena such as climate and subjects like sustainability in Chapter 8. His historical account of what we now call 'sustainability' makes a clear change from an implicit towards an explicit future. That is to say, thinking about the future of our planet is also a tool to set the agenda and to try to make the necessary adjustments in how we deal with life on earth. That the value of looking to future is only measured by assessing the accuracy of predictions is illustrated by May by stressing that many authors of alarming environmental reports hope and pray that their predictions will not come true.

How a rather modern science such as the science of management copes with the future is the topic of Chapter 9, written by dr. Patrick van der Duin and dr. Erik den Hartigh (both Delft University of Technology, The Netherlands). The focus is on strategy and innovation. They very much view the future from the perspective of the manager or entrepreneur, who not only thinks about the future in terms of strategy and innovation but who has an underlying attitude towards his business environment. His view on his business and the assumed or believed capability of shaping 'his' business are directly influencing his attitude toward strategy and innovation and thereby to the future.

Which role the future plays in astronomy, perhaps the mother of all (natural) sciences, is described by dr. Dap Hartmann (Delft University of Technology, The Netherlands) in Chapter 10. Based on the observation that astronomers are looking at the past, he shows that for astronomy and physics these historical data are vital in predicting the future. Also, he shows that the extraordinary scales of distance and time with regard to astronomical topics gives us a different perspective on time as well as the future.

Last but not least, prof.dr. Alan Porter (GeorgiaTech University, U.S.A.) assesses in Chapter 11 which role currently the future plays in different sciences. By using very modern bibliometric techniques, he 'measures' how large a role 'futures research' (Porter uses very much synonyms) plays in scientific literature. It shows that publications in the field of 'Future-oriented Technology Analysis' (as Porter calls it) are increasing, that the U.S produces the most publications in this field, and that FTA literature "centers in the 'Quantitative Applications'" area.

Finally I would like to express my gratitude to everyone who has contributed to this publication. First of all, all authors who have done much effort to think about their subject. I hope for them the writing process was also a kind of 'sentimental journey' since many wrote about their educational background from which they made a shift to the world of futures research. Second, my gratitudes to those who have reveiwed one or more chapters of this book: Roland Ortt, Enid Mante, Cees van Beers, Joris de Rooij, Adrienne van den Boogaard, Jacco Quist, Frida de Jong, Mario Jacobs, Thijs van Erve, Guus Berkhout, Marc Zegveld, Claire Stolwijk, Simone Arnaldi, and Roos Bonnier.
Also, I would like to thank Wiebe de Jager of Eburon Academic Publishers who believed very strongly in this book and who has done a lot to make its publication possible. Many thanks as well to Gert Stronkhorst for correcting and editing some of the chapters in this book. I also would like to thank the section Technology, Strategy, Technology and Entrepreneurship of the faculty Technology, Policy and Management of Delft University of Technology and Robert van Oirschot of consultancy company Alares (the Hague, the Netherlands) for supporting this book financially. A final thanks to prof.dr. Wim van der Donk and prof.dr. Anton Hemerijck of the Netherlands' Scientific Council for Government Policy for their financial contribution..

References

Barker, J. 1996, *Paradigms. The business of discovering the future,* Harper Business, New York.

Bell, W. & J.A. Mau 1971, *The sociology of the future,* Russell Sage Foundation, New York.

Van der Duin, P.A. 2006, *Qualitative futures research for innovation,* Eburon Academic Publishers, Delft.

2 Inside the foresight mind

Peter Hayward

1 Introduction

How is it that individuals conceive of the future? What does a survey of relevant psychological theory bring to an understanding of futures thought? The future is not something that can be empirically located and quantified. The future cannot be reduced to a mathematical proposition or formula. The future is not a 'received view' but is, instead, a 'constructed view'. To some this would mean that the study of the future is not scientific; that the only knowledge of any significance is that which is derived from the use of objective measures. Yet thinking about the future is not an abstract theory but is a biological and cultural fact. We regard foresightful behaviours that defer immediate gratification as healthy or even wise. We marvel at animals like beavers and squirrels that display future-orientated behaviours. The English language has preserved truisms that embody future orientation – "look before you leap" and "a stich in time saves nine." Thus we have an objective process that deals with a subjective construction. Kuhn (1970) framed the human knowledge quest as one understood as a series of 'revolutions' in paradigms, or exemplars, of knowledge injunctions. Yet the paradigms themselves did not eventuate outside the knowledge quest but were brought forth specifically by the actions of deciding that a particular problem should be solved. Particularly:

> *since no paradigm ever solves all the problems it defines and since no two paradigms leave the same problems unsolved, paradigm debates always involve the question: Which problems is it more significant to have solved? Like the issue of competing standards, that question of values can be answered only in terms of the criteria that lie outside of normal science altogether, and it is that recourse to external criteria that makes paradigm debates revolutionary (Kuhn 1970, p. 14).*

Psychology itself has undergone a number of paradigmatic revolutions as the significant question of how human thought eventuates has been debated. This chapter will employ those debates in order to uncover how psychology can aid our understanding of how futures thought might eventuate, be studied and propagated.

2 Conscious animals

Evolutionary biology's contribution to our understanding of futures thought arises from the consideration of the brain and consciousness. Evolutionary biology's knowledge of brain structure and function continues to expand and it does provide considerable insight into the nature of futures thought. Brain structure is complex and will continue to challenge researchers but that research is simplified by the fact that the brain is a physical object that has location an can be measured. Brain structure and function is being progressively defined and agreed by researchers. The same cannot be said for the study of consciousness. Consciousness is non-locatable and defies objective measurement and so tends towards a plurality of definitions with little agreement amongst researchers. This section will summarise the agreed points of research into brain structure and function and how these relate to futures thought. It will not, however, summarise the agreed points of research into consciousness and their relation to futures thought but it will represent some views on consciousness that, in the opinion of the author, offer utility to the study of future thought.

Edelman (1992) posited a theory of evolutionary and biological development distinguishing two forms of consciousness, a primary consciousness (a creature present) and a 'higher-order' consciousness. His theory is premised upon two neurological capacities, *value determination* and *categorisation* and how they enable individual consciousness. *Value* is a property of the brain stem in conjunction with the limbic (hedonic) system. These two systems manage the body and its functions. They are directed not towards stimuli in the external environment, but rather, the regulation of the body in order to maintain the conditions for life by way of evolutionarily selected value patterns. *Categorisation* is a property of the thalamocortical system (the thalamus and the cortex). This system receives its information from outside of the body and it creates detailed mappings of the signals received through the senses. The thalamocortical system, the brain stem and the limbic system were linked at some evolutionary point thus allowing the categorisations to be mediated by evolutionary values. The mapping of the sensory stimuli produces a 'scene' and the value-memory highlights the salient events in that scene. Certain events could be identified which had a higher relative value than others. Behaviours could then be adapted to better suit the value patterns. This combined system has a clear evolutionary advantage as it served adaptive behaviour that better suited the environment. This is what Edelman calls *primary consciousness*. This form of consciousness, however, still lacks a complex notion of self, cannot model the past or future and to a significant extent is only able to correlate external perceptions. Edelman (1992, p. 122) describes an animal with only primary consciousness as seeing 'the room the way a beam of light illuminates it. Only that which is in the beam is explicitly in the remembered present; all else is in darkness.' Damasio (1999) describes this type of consciousness as *sentience*, 'providing the organism with a sense of self about one moment – now – and about only once place – here...it does not illuminate the future.' Nevertheless primary consciousness is necessary for the development of higher-order consciousness.

The key to the evolution of higher-order consciousness is *concept formulation*. This is the ability to control behaviour by means of stimuli not arising from external perception. In comparison to perceptual categorisation which is unconscious and externally focussed, conceptual categorisation is conscious and internally focussed. Reading (2004, p. 63) argues that this ability to formulate concepts that can operate as if they were perceptual stimuli distinguishes humans from other animals. 'Other animals do not possess the neuronal capability for constructing a symbolic model of the universe – and are thus limited to processing present-orientated experiences.'

The evolution of linguistic capabilities in Homo Sapiens is an obviously critical step in the evolution to higher-order consciousness. Articulated sounds and their symbolic meaning required a new, or at least dramatically enhanced, form of conceptual memory – e.g. not only the memory of the process but the memory of the sound – as well as memory itself. Further, the conceptual categorisations of linguistics were linked to value-category memory so as to ensure that communication too would be categorised with evolutionary value-patterns. This adaptation of conceptual categorisation would give further evolutionary fitness to Homo Sapiens in continuing to adapt behaviour to environmental complexity. As the semantic lexicon increases, syntax emerges as a new form of conceptual categorisation. Symbolic meaning can also be conveyed through the plasticity of the syntactical items themselves. Thus a sense of self and its link to primary consciousness through value-category memory can emerge. With a sense of self can also arise a sense of a world which the self occupies, as well as a self which changes in that world. The 'ever-present' present of primary consciousness is now expanded by a memory of a present recently removed – the past. With a memory of past comes the anticipation of a present not yet arrived – the future. These concepts arise within an interaction with other members of the same species, through social transmission and learning, all of which is still mediated by the evolutionary value patterns.

> *The freeing of parts of conscious thought from the constraints of an immediate present and the increased richness of social communication allow for the anticipation of future states and for planned behaviour. With that ability come the abilities to model the world, to make explicit comparisons and to weigh outcomes; through such comparisons comes the possibility of reorganising plans. Obviously these capabilities have adaptive value. The history of humanity since the evolution of hunter-gatherers speaks to the adaptive and maladaptive properties of the only species with fully developed higher-order consciousness (Edelman 1992, p. 133).*

The future thinking capacity of higher-order consciousness is observed in circumstances where disease or injury has damaged the relevant parts of the brain. 'The frontal lobe empowers healthy human adults with the capacity to consider the self's extended existence throughout time...people with frontal lobe damage are described by those who study them as 'bound to present stimuli' or 'locked into immediate space and time' or as 'displaying a tendency to temporal concreteness'...they live in a world without later' (Gilbert 2006, p. 14). A healthy brain is a necessary for higher-order consciousness and without it the capacity for futures thought can be lost but neuropsychology alone cannot explain how consciousness leads to the development of futures thinking. Psychology has researched the individual capacity for futures thought and it is this area upon which the remainder of this chapter shall concentrate.

3 The psychological understanding of futures thought

Psychology regards futures thought as the broad capacity to conceptualise an idea of the future. Psychology has commonly woven together individual future thought with ideas of individual motivation.

> *The ability to foresee and anticipate, to make plans for and to organize future possibilities represents one of the most outstanding traits of man. This orientation ahead is more than an expression of the continuous effort towards a better adaptation between man and the world. The restless striving is just more than*

a drive for competence and display...Man's unique ability to conceptualize time enables him to anticipate and organize future possibilities and thereby to bring effects of future time into the psychological present (Gjesme 1983a, p. 347).

To many, the taking of action was an expression of future thought. Almost all action could be seen to contain a future element. 'On the behavioral level the object needed is something to strive for or achieve and this constitutes the behavioral future...the future is primarily motivation space' (Gjesme 1983*b*, p. 446). Thus to study why individuals act, how they are motivated, was also to study how individuals think about the future. Much of the early psychological theory about individual action was founded upon the view that human behaviour was shaped and controlled automatically and mechanically by environmental stimuli (Bandura 2001, p. 2). This theory's two key assumptions were that:

- only material events could be causal, and
- only entities that were directly, externally perceivable could be admitted into the realm of science.

Motivational psychologists kept to these assumptions by externalizing or materializing their key concepts. Skinner externalized motivation by attributing it to reinforcers while drive-reduction theorists, like Hull, kept motivation inside the organism but attributed it to strictly physiological mechanisms (Locke 1996, p. 117). Yet this type of theorising lead to a research dead-end as it had to ignore time as a concept. 'Behavioural research was hard put to include time as a variable, in that it is closely linked to the phenomenal aspects of our experience of time and is not a physiological factor' (Fraise 1984, p. 3).

Beginning in the late 1960s the positivist paradigm in psychology began to fall apart for a number of reasons. First it had lost support in philosophy. Second the materialist approach did not work. Human action cannot, in fact, be understood by looking at man only from the outside or only at his internal physiology. The recognition of these facts ushered in the 'cognitive revolution' in psychology (Locke 1996, p. 117).

Instead of studying motivation as if it were impersonal, instinctual and even unconscious, psychologists began to focus on self-relevant thoughts as behavioural mediators in order to personalise motivation (Leonardi, Syngollitou & Kiosseoglou 1998, p. 153). The centrality of consciousness to both motivation generally, and future thought specifically, was admitted. 'Without a phenomenal and functional consciousness people are essentially higher-level automatons undergoing actions devoid of any subjectivity or conscious control' (Bandura 2001, p. 3). On reflection, without this change occurring, it is inconceivable that the conception of future time in individual thought could be usefully researched at all. It is the uniquely human capacity of contrasting present reality with possible futures that establishes a behavioural commitment to act (Oettingen 2000, p. 114). 'Forethoughtful, generative and reflective capabilities are, therefore, vital ones for survival and human progress' (Bandura 2001, p. 3). Cognitive psychology provided a basis upon which the understanding of how future thought eventuated in individuals could be studied.

3.1 Cognitivism and future time perspective

> *"What is specific for human beings is that we have, with cognitive means, increased the range and the precision of usable natural rules so much that we can extend our inferences into the distant past or distant future" (Toda 1983, p. 353).*

The essence of cognitivism is flexibility and adaptation. An individual's mind must be able to depict stable internal representations of the environments in which it operates. It is the representation of order and predictability that gives much of the confidence needed to act. Such representation is an example of an integrative operation between meaning and experience that is necessary to maintain organic autonomy. Conversely to guide behaviour in a truly flexible manner, the mind must also be able to respond to the presence of the unexpected. Macrae & Bodenhausen (2000) argued that an adaptive mind is one that enables its owner to override automated action plans and produce novel behavioural outputs as and when those responses are required.

> *What was new in the cognitive system operation was that it went beyond building object representations and started to build a world representation. Once such a world representation is obtained, it aids the cognitive system operations by providing a reliable context for planning one's future behaviour (Toda 1983, p. 361).*

Cognitivism believed that the mind placed objects into an environmental context that permitted both a sense of balanced autonomy and also flexibility in behaviours. 'The ability to bring anticipated outcomes to bear on current activities promotes foresightful behaviours. It enables people to transcend the dictates of their immediate environment and shape and regulate the present to fit a desired future' (Bandura 2001, p. 7). Behaviouralism gives no consideration to cognitive processes. Cognitivism placed the mind in the process in order to acknowledge human will. That representation of human cognitive functioning, however, was regarded as 'cold cognition' as it assumed that the mind operated from 'rational choice': the idea of the mind in total control of the emotions. 'Decision makers are supposed to rationally calculate for all possible courses of actions the utility of each possible outcome, and weigh the utilities with the probability that each outcome will occur' (Zeelenberg 1999, p. 2).

Yet such a model of cognition flies in the face of the fact that thoughts can be felt as, more or less, positive or threatening and the finding that emotions do affect decisions made. One study found affect to have a greater influence on judgements in direct correlation with the elaboration of the thinking used. It was theorised that affect has less influence when situations are more familiar and so less elaborate thinking is needed (Ciarrochi, Forgas & Mayer 2001, p. 53). In those situations habitual thinking is activated and affective influence is minimised, whereas when the situation is abnormal and the thinking is elaborated then affective influence is maximised. As futures thinking tends to be practiced in non-habitual situations, and hence its approach would likely employ elaborated thinking, it follows that the affective influence on futures thought would be significant. Affect could colour and even override cognition of the future. The central point was not whether affect was a dimension of future orientation, it was, but whether affect could be consciously shaped by individual cognition. Could an individual, affect notwithstanding, make decisions that led towards healthy outcomes? Did future thought play a role in those decisions?

3.2 *The clinical interest in futures thinking*

There is a body of research into future orientation in individuals that has a strong clinical interest.

> *Most if not all, research on future time perspective has been guided by clinical interest, following the general assumption that an extended future time perspective leads to a well-adapted and psychologically healthy personality. To bring the future into the present, the individual has to have, or develop, the capacity to plan his or her activities. Planning is facilitated by developing proximate goals that intervene between one's present state and the desired ultimate, distant goal (Seijts 1998, p. 2).*

On the first reading, this research 'gives the impression that a well-structured and extended future orientation is a characteristic of a well-adapted personality...and allows for activities that are highly valued in our culture' (Trommsdorff 1983, p. 381). The underlying logic beneath this impression is that:

Future orientation > Goal setting > Planning > Positive Life Outcomes > Healthy Personality

The research into extrinsic and intrinsic goals supports this logic. Intrinsic goals are orientated towards growth relevant, inherently satisfying activities, whereas extrinsic goals are more focussed in the attainment of external rewards and praise (Husman & Lens 1999, p. 113).

> *As predicted, the extent to which strivings to bring about intrinsic futures was correlated with more positive well-being outcomes than was the extent to which strivings to bring about extrinsic futures. In addition, participants whose strivings helped bring about intrinsic futures were more likely to be autonomously orientated and to engage in meaningful activities, whereas participants whose strivings helped extrinsic futures were more likely to be control orientated and to engage in distracting activities (Sheldon & Kasser 1995, p. 540).*

This type of research demonstrated a strong correlation between life conditions that supported 'intrinsic futures thinking' and a healthy personality. Other research, however, failed to find the same strong correlation.

Other clinical interest research sought to determine the significant correlates between other personal variables and future thought. The relation between variables such as anxiety, dogmatism, internal control, achievement motivation, schizophrenia, delay of gratification, consuming behaviour and future thought were all studied. 'In some of these studies, such relations are significant in the expected direction, in other studies, relations between future orientation and behaviour are mediated by other variables, or do not emerge at all' (Trommsdorff 1983, p. 385). In short: 'investigations in this field can form almost any conclusion you prefer to draw. The inconclusiveness of empirical findings is almost complete' (Gjesme 1983*b*, p. 445). The inability of the clinical research to clearly determine the individual personality correlates saw the conclusion reached that future thought is:

> *a multi-dimensional cognitive-motivational construct: cognitive schemata on the subjective future may be differentiated according to their temporal and causal structure and underlying subjective judgements of future events as more or less probable; the motivational and affective quality of the subjective future may be differentiated according to the affective quality of the future (as more or less positive or threatening) and its specific thematic content (goals)...The nature of systemic relations between future orientation and other personal variables is very unclear (Trommsdorff 1983, p. 384).*

The clinical research that sought to demonstrate the assumed relationship between future thought and 'highly valued actions' was largely inconclusive. No single personality factor could account for future thought in individuals. Research that studied future thought, however, through environmental and social factors uncovered a different hypothesis.

Klineberg (1967) found that an extended future orientation did not necessarily indicate a well-adapted personality. Well-adjusted adolescents had a shorter but more realistic future time perspective than maladjusted adolescents who were more influenced by wishful-thinking tendencies. Other research demonstrated that delinquents realistically structured their future more in terms of fear than hope. This made sense given that in the life conditions of a delinquent most of the problems to be coped with will be found in the near future while the distant future could not be expected to be a positive one (Trommsdorff 1983, p. 386). That socially disadvantaged groups would realistically avoid anticipations of the future, as they most probably are associated with pessimistic outcomes, was also supported by Shannon (1975). While Anglo-American adolescents developed a more extended future orientation, Indian and Mexican Americans did not develop such an extension (Shannon 1975, p. 114). The similar conclusion drawn was that these minority groups had learned about the future difficulties that they would face in attaining the same goals of the majority group. Another study found that high and prolonged deprivation shortened the future time orientation in individuals. Importantly this study found that the deprivation of physio-economic comforts did not have a significant effect while the deprivation of the experiential dimension significantly and adversely affected future orientation. 'The present findings show that in spite of highly deprived physio-economic conditions, one can learn to be future orientated and plan for future opportunities provided one receives adequate parenting, emotional, motivational and educational experiences' (Agarwal, Tripathi & Srivastava 1983, p. 377). In summary, the clinically orientated research found that environment and experience significantly shaped future thought. Future thought is shaped by the environment while the behaviours that arise from that mode of futures thought seek environmental assimilation.

Deprivation in the experiential, rather than physio-economic, domain was most significant on the development of future thought. Given benign environmental and experiential conditions, aspects of personality did play a role, however, no single personality factor could reliably explain future time development in individuals. The conclusion drawn was that future orientation is a multi-dimensional cognitive construct but the nature of how that construct operates was not clear. Beyond these findings what would further advance understanding is ideas of how this interaction between environment, experience and future thought eventuated. The theories of developmental psychology are useful here.

3.3 Developmental psychology and futures thinking

> *Empirical evidence suggests that future time perspective is a cognitive structure rather than a stable disposition. First the capacity to experience time, and estimate it, has been found to be a gradually developing characteristic. The ability to extend the idea of time into both the past and the future continues to develop with age (Seijts 1998, p. 3).*

What is the dynamic of this developmental process? What is it that is developing? What the clinical studies were observing were the behavioural manifestations of futures thought. Was thinking past, present or future focussed? If future focussed, how far did it extend into the future? How coherent and organised was the future thought? While development in those behavioural manifestations can be observed, that does not get closer to why and how the future thought developmental dynamic is eventuating. That question can be usefully addressed through the closer examination of the process of how environment and experience impacts upon cognition.

> *To know is to assimilate reality into systems of transformations. To know is to transform reality in order to understand how a certain state is brought about... Knowing reality means constructing systems of transformations that correspond, more or less adequately, to reality. The transformational nature of which knowledge consists are not copies of the transformations in reality; they are simply possible isomorphic models among which experience can enable us to choose. Knowledge, then, is a system of transformations that become progressively adequate (Piaget 1970, p. 15).*

It was Jean Piaget's study of the development of cognition in children that uncovered the essential point that thinking is a self-created structure that seeks to assimilate reality. This is the symbolic transformation of organic experience that the philosophy of time theorised about. Cognition does not develop from deficient forms to efficient forms but rather each structure of cognition provides an adequate assimilation of reality at that point in time. Each structure of thought maintains the coordination of experience and environment that is necessary to ensure psychological autonomy. Subsequent structures of thought assimilate a greater range of experiences and environments and thereby become more adequate. What Piaget determined was that the developmental patterning of these structures in children were isomorphic. The children were not copying the thinking of others. Instead they were developing their own assimilative structures and the nature of these patterns of thinking uncovered an underlying developmental structure in individual cognition. Considering future thought from within developmental psychology provides a 'more adequate' understanding of the dynamic of the process.

Concepts of time develop in a way similar to concepts of space, volume or matter; e.g. the child derives concepts of time from relationships among more simple concepts such as work and power. Accordingly one has to assume that formal operational intelligence enables the child to anticipate consequences and to think in terms of future possibilities instead of simple prolongation of actions or operations... [Klineberg] shows that, under normal conditions of cognitive development, older children (between 14 and 16 years) are able to anticipate future developments and extend their future time perspective realistically. Unsuccessful persons fail to develop an extended future orientation since – from a realistic point of view – the future will bring them undesired developments. Klineberg's study supports Piaget's cognitive theory on the development of time perspective; furthermore it shows very clearly that besides endogenous factors, such as cognitive maturation, the development of future orientation is influenced by situational factors such as social and economic conditions (Trommsdorff 1983, p. 389-90).

Here again the centrality of environment and experience on future thought is restated. Given, at least, benign life conditions then future thought should develop as cognition matures. If the life conditions allow, then structures of thought can increase the adequacy of environmental and experience assimilation. What is still missing, however, is an understanding of what does the 'assimilating'. The idea of a cognising 'self' is relevant here.

3.4 The self and futures thought

What is meant by the term 'self'?

The self regulates behaviour, sets goals and expectations, motivates performance to meet these goals, monitors performance on different tasks and evaluates whether performance fulfilled the goals. Motivation, in the light of this theory, is seen not as a generalised disposition or a set of specific goals but as a reflection on what individuals hope to accomplish with their life and the kind of people they would like to become. Possible selves are defined as conceptions of the self in future states. They are thought to derive from representations of the self in the past and to include representations of the self in the future. They are considered different and separable from the now of current selves but intimately connected to them. They are regarded as the cognitive manifestations of enduring aspirations and motives (Leonardi, Syngollitou & Kiosseoglou 1998, p. 154).

This journey along the pathway of futures thought has now traveled a considerable distance from the deterministic presumptions that underpinned behaviouralism. The clinical research into future thought concentrated upon the study of behaviour, particularly maladaptive behaviours, as that interest was directed towards the encouragement of what was healthy and the discouragement of what was harmful. While laudable, adopting a positivist research paradigm towards the study of futures thought ignored the point that the future can only be subjectively understood. The future is a symbolic construction of the individual mind. While these constructions do manifest in observable behaviours, that does not change the essential subjective ground and nature of the thinking. Furthermore, while futures thought can be seen as an operation of cognition it does not arise solely from cognition. Rather, the idea of a conative and agentic self that co-creates the concept of 'future' in order to find adequate assimilation of their experiences and environment is fundamental to its understanding. A sense of self develops through environment interaction and, in turn, this moderates futures thought and futures behaviours.

> *Although the dominant paradigm in psychology is skewed towards a deterministic perspective, there is a rich history and strong support for a phenomenological perspective. For example, the work of James (1890) was largely based on this perspective, from which he identified a number of "I" processes, including awareness of one's (a) personal continuity over time, (b) distinctiveness, (c) agency over life events, and (d) volition (McCombs & Marzano 1990, p. 53).*

The individual assimilation of experience and environment is aided by the development of a healthy 'ego' or self-sense. One attribute of healthy ego or self-sense is the belief of an internal-external control dynamic. Without an individual sense of internal control 'events are essentially unpredictable because they depend on external forces, whereas in the case of internal control they are predictable for they depend on what the person believes in his own initiative. This involves a certain confidence in planning and visualizing the future' (Rabin 1978, p. 299). The individual sense of their potential agency over environment and experience arises here. 'The individual's awareness of the self as an agent is a key factor in motivating behaviour' (Leonardi, Syngollitou & Kiosseoglou 1998, p. 153). Without the sense of the 'I' who acts, individual behaviour is reduced to the mere reflex or reaction to stimulus or drive that was promulgated by behaviouralism.

> *Defining the ego is difficult in the same way that defining life is. Air and water are not living beings. When one drinks water or breathes air, at what point does it become part of a living object?...If we think of life as being a process of interchange with the environment, the question loses point. There is no problem...Piaget uses*

the term mobile equilibrium – the more mobile, the more stable. The striving to master, to integrate, to make sense of experience is not one ego function among many but the essence of ego (Loevinger 1976, p. 58-9).

The future, therefore, is the motivational space of the ego or the cognising self. It is the place where the self, as agent, anticipates the outcomes of its present-based behaviours. As foresight is both the capacity of bringing a consideration of the future into the present and it is also a range of actions based upon that consideration then the idea of a cognising self draws the psychological research into futures thought close to the question posed at the start of this chapter. Such a motivational space, however, is not necessarily a consciously accessible space. The future only fully becomes an accessible space for thinking 'about' through the advent of metacognition.

3.5 Metacognition and its role in creating agentic futures thought

Metacognition is the capacity of the individual to understand their thinking, 'as an operation itself', and thus to see themselves as not only having agency over their behaviours but also agency over the thinking that precedes behaviour.

People are not only agents of action but also self-examiners of their own functioning. The metacognitive capability to reflect upon oneself and the adequacy of one's thoughts and actions is another distinctly core human feature of agency. Through reflective self-consciousness, people evaluate their motivation, values and meaning of their life pursuits...Among the mechanisms of personal agency none is more central or pervasive than people's belief in their capability to exercise some measure of control over their own functioning and over environmental events. Efficacy beliefs are the foundation of human agency (Bandura 2001, p. 10).

Metacognition, in effect, allows for the assimilation of not only consciousness, but also self-consciousness with experience and environment. The operations of metacognition can also be reflected upon 'metacognitively', thereby dynamically accelerating the assimilative process.

Thus metacognitive understanding is not a process of intellectually constructing a schema that includes the role of self, but is an ongoing process of progressively deeper insights or realizations that, in turn, lead to an awareness, or conscious understanding of self as agent...the metacognitive, cognitive, and affective systems are more accurately described as subsystems operating in support of the self. They reside functionally under the control of the self as agent – under the control of the 'I' – the experience of being and volitional agency (McCombs & Marzano 1990, p. 54).

Metacognition can be described as the cognitive process of the 'self-as-agent'. The ability to bring a level of understanding above the operations of instinct, emotion and thought create the possibility of consciously 'regulating the use of these knowledge structures in support of personal goals, intentions and choices. A realization of self as agent automatically leads to self-determined purposefulness' (McCombs & Marzano 1990, p. 55).
Gilbert (2006) explains our foresight capacity as a flawed process that:

allows us to mentally transport ourselves into future circumstances and then ask ourself how it feels to be there...Our ability to project ourselves forward in time and experience events before they happen enables us to learn from mistakes without making them and to evaluate actions without taking them...as impressive as it is, our ability to simulate future selves and future circumstances is by no means perfect. When we imagine future circumstances we fill in details that won't come to pass and leave out details that will. When we imagine future feelings, we find it impossible to ignore what we are feeling now and impossible to recognize how we will feel about things that happen later (Gilbert 2006, p. 238).

For Gilbert it is our metacognitive abilities that allow us to understand how foresight is a 'fragile talent' and that we can use more, or less, wisely once we understand what it does well and what it does poorly.

3.6 The 'fallacy' of thinking about thought

A contrarian argument to what has previously been outlined should also be put, if not only for balance, then for the possibility that our understanding of consciousness and time could be very off-beam. McGinn (2006) argues that "there are nontrivial limits on what human beings can come to grasp (p. 109). He says that our entire conceptual scheme is "shot through with spatial notions" (p. 110) and so we could mistake our ability to properly understand the nature of consciousness.

Our knowledge constitutes a kind of 'best fit' between our cognitive structure and its objective world; and it fits better in some domains than others. The mind is an area of relatively poor fit. Consciousness occurs in objective reality in a perfectly naturalistic way; we just have no access to its real inner constitution ... Since our bodies are extended objects in space, and since the fate of these bodies is crucial to our reproductive prospects, we need a guidance system in our heads that will enable us to navigate the right trajectory through space, avoiding some objects (predators, poisons, precipices) while steering us close to others (friends, food, feather beds). Thus our space-representing faculties have a quite specific set of goals that by no means coincide with solving the deep ontological problems surrounding consciousness and space (McGinn 2006 p. 113).

It follows then that we may suffer the illusion that we understand consciousness much better than we do. Such an argument does not, necessarily, negate all that has been previously said about the psychological nature of futures thought. It may be that those conclusions do contain veracity. What McGinn's argument does do is suggest that we cannot easily dismiss understandings of consciousness and the future that do not correspond comfortably to our naturalistic notions of the objective world.

One such 'unnatural' notion of consciousness and future thought is 'synchronicity'. Jung (1963) posited that his concept of archetypes and the collective unconscious included the notion of 'meaningful co-incidence' or 'acausal parallelism' (p. 342) – the idea that psychic and physical phenomena can be causally related notwithstanding that they do not share temporality. The Princeton Engineering Anomalies Research Laboratory engaged in over twenty years of research into whether a interactions between mind and matter could be proven.

Bradley (2007) has theorised and demonstrated that the future can be intuitively foretold by means of non-localised communication that operates precognitively, this time via the combination of bio-emotional energy and quantum holography. Ideas such as 'non-localised communication' or 'parallelism' do challenge our naturalistic sense of how we comprehend our own reality but they are nevertheless still making a disciplinary contribution to the possible understandings of consciousness and psychology.

4 Conclusion

Notwithstanding that, in the view of this author, psychological considerations do offer much to our understanding of futures thought, the sense that one has is that the ambition of psychology has been foreshortened. Mind is a troublesome research subject while brain research is less problematic. If research into future thought is reported nowadays then it is likely to be reporting on which elements of the brain are 'triggered' when individuals think about the future (Szpunar 2007). Contemporary research shies away from research of the phenomenological nature of futures thought, probably because that research stance does not produce objective evidence. This does not mean that we have returned to the crude behaviouralism of the 50s and 60s that disregarded individual subjectivity and regarded individuals as automatons who responded to. Nevertheless the drift of psychological understandings of futures thought via brain structure and function does seem eerily familiar of that old research 'dead end'.

Brain structure and function is fundamental to futures thought and is not disputed. Functional development, at least amongst paediatric researchers, is broadly accepted. With advances in brain imaging technology the objective correlates of that developmental process will likely be observed. What is missing, and shows no current sign of being addressed by mainstream psychological research, is ongoing adult cognitive development and its impact on psychological capacities, like futures thought. It feels unsatisfactory to have to rely on theories of adult psychological development that are 30-40 years old in order to explain how cognitive processes like futures thinking can become elaborated in terms of temporal frame and depth of analysis. It feels unsatisfactory because those theories of adult development are largely disregarded by current psychology.

As Kuhn (1970, p. 14) proffered "no paradigm ever solves all the problems it defines" and the significance of problems to be solved are a "question of values [that] can be answered only in terms of the criteria that lie outside of normal science altogether". If psychological theories of 'the mind' are no longer considered significant to psychology then perhaps they could be regarded significant in the study of futures thought in adults? At least Kuhn does permit that question to be resolved by reference to things outside science. If the development of futures thinking in adults that encompassed concepts like 'non-violent futures', moral futures', 'concern for future generations' or 'zero environmental footprint futures' is valued then psychology does offer us useful theories and frameworks to do just that.

References

Agarwal, A., Tripathi, K. & Srivastava, M. 1983, 'Social roots and psychological implications of time perspective', *International Journal of Psychology*, vol. 18, no. 5, pp. 367-80.

Bandura, A. 2001, 'Social cognitive theory: An agentic perspective', *Annual Review of Psychology*, vol. 52, no. 1, pp. 1-26.

Bradley, RT 2007, 'The psychophysiology of intuition: a quantum-holographic theory of nonlocal communication', *World Futures*, vol. 63, no. 2, pp. 61-97.

Ciarrochi, J., Forgas, J. P. & Mayer, J. D. 2001, *Emotional Intelligence in Everyday Life*, Psychology Press, Philadelphia, USA.

Edelman, G. 1992, *Bright air, brilliant fire*, BasicBooks, USA.

Fraise, P. 1984, 'Percpetion and estimation of time', *Annual Review of Psychology*, vol. 35, no. 1, pp. 1-36.

Gilbert, D. 2006, *Stumbling on Happiness*, HarperPress, London, UK.

Gjesme, T. 1983*a*, 'Introduction: an inquiry into the concept of future orientation', *International Journal of Psychology*, vol. 18, no. 5, pp. 347-50.

Gjesme, T. 1983*b*, 'On the concept of future time orientation: considerations of some functions and measurements implications', *International Journal of Psychology*, vol. 18, no. 5, pp. 443-61.

Husman, J. & Lens, W. 1999, 'The role of the future in student motivation', *Educational Psychologist*, vol. 34, no. 2, pp. 113-125.

Jung, C.G. 1963, *Memories, Dreams, Reflections*, Collins and Routledge, London.

Klineberg, S.L. 1967, 'Changes in outlook of the future between childhood and adolesence', *Journal of Personality and Social Psychology*, vol. 7, pp. 185-93.

Kuhn, T.S. 1970, *The structure of scientific revolutions*, Vol. 2 of *International Encyclopedia of Unified Science*, 2 edn, The University of Chicago Press, Chicago, USA.

Leonardi, A., Syngollitou, E. & Kiosseoglou, G. 1998, 'Academic achievement, motivation and future selves', *Educational Studies*, vol. 24, no. 2, pp. 153-63.

Locke, E. 1996, 'Motivation through conscious goal setting', *Applied and Preventive Psychology*, vol. 5, pp. 117-24.

Loevinger, J. 1976, *Ego Development*, Jossey-Bass, San Francisco, USA.

Macrae, N. & Bodenhausen, G. 2000, 'Social cognition: Thinking categorically about others', *Annual Review of Psychology*, vol. 51, no. 1, pp. 93-120.

McCombs, B. & Marzano, R. 1990, 'Putting the self in self-regulated learning: the self as agent in integrating will and skill', *Educational Psychologist,* vol. 25, no. 1, pp. 51-69.

McGinn, C. 2006, *Consciousness and its Objects*, Oxford University Press, Oxford

Oettingen, G. 2000, 'Expectancy effects on behaviour depend on self-regulatory thought', *Social Cognition,* vol. 18, no. 2, pp. 101-129.

Piaget, J. 1970, *Genetic Epistemology*, Colombia University Press, New York, USA.

Rabin, A.I. 1978, *Future Time Perspective and Ego Strength,* Vol. 3 of *The Study of Time,* ed. J. T. Fraser, N. Lawrence & D. Park, Springer-Verlag, New York, USA.

Seijts, G.H. 1998, 'The importance of future time perspective in theories of work motivation', *Journal of Psychology,* vol. 132, pp. 154-168.

Shannon, L. 1975, 'Development of time perspective in three cultural groups', *Developmental Psychology,* vol. 11, pp. 114-5.

Sheldon, K.M. & Kasser, T. 1995, 'Coherence and congruence: two aspects of personality integration', *Journal of Personality and Social Psychology,* vol. 68, pp. 531-543.

Szpunar, K., Watson, J. & McDermott, K. 2007, 'Neural substrates of envisioning the future', Proceedings of the National Academy of Sciences of the United States of America, vol. 104, pp. 642-7.

Toda, M. 1983, 'Future time perspective and human cognition: an evolutional view', *International Journal of Psychology,* vol. 18, no. 5, pp. 351-65.

Trommsdorff, G. 1983, 'Future orientation and socialization', *International Journal of Psychology,* vol. 18, no. 5, pp. 381-406.

Zeelenberg, M. 1999, 'The use of crying over spilt milk: a note on the rationality and functionality of regret', *Philosophical Psychology,* vol. 12, no. 3, pp. 325-41.

3 Futures studies from a human and social perspective[1]

Eleonora Barbieri Masini

1 Introduction

This essay will focus on how the social sciences have addressed future perspectives which, in their turn, challenge social sciences as a whole and each of them individually, including sociology. The challenge arises from the continuously changing social environment. It is probably a truism to say that changes at all levels of society are growing not only stronger but also more closely interrelated, and that they have a global character.

This essay illustrates how a human and social perspective has emerged in Futures Studies (FS) during the fifty-years long history of this research field. This human and social emphasis is twofold: on the one hand, FS have engaged in the anticipation and design of societal futures; on the other, sociological concepts, theories, and methods have enriched and fertilised the futures thinkers' toolbox and their conceptual repertoire.

Despite the link, the relation between FS and sociology has been often neglected or disregarded (Barbieri Masini 1999, pp. 325-31). The relation existed in the past and exists now, and hence it cannot be discarded simply as irrelevant for the future. Acknowledgement of this relevance has two major implications: on the one hand, futures researchers are obliged to frame temporarily-bounded and future-oriented concepts in a rich sociological tradition, which is far older than futures research itself; on the other, the sociological understanding of time and the future is an essential term of reference for dealing with human and social futures studies.

1 I wish to thank Simone Arnaldi for his help in editing the sociological part in this essay.

Though an account on the sociological understanding of time is far beyond the scope of this essay, it may be useful to present some brief thoughts on this topic, as an introduction to the next sections, which are focused on futures studies.

Among the classics, for instance, time-oriented representations have an important place in Durkheim's theorising on the elementary forms of religious life (1912/2005) and Weber's account of Protestant ethics and capitalism (1905/1991). As I explain below in this essay, the concept of representations of the future subsequently influenced several sociologists and social scientists who in the 1960s and 1970s more explicitly engaged with the relation between sociology and futures studies, such as Fred Polak, who extensively commented on Max Weber's *Protestant ethics and the spirit of capitalism* in his theory of the relations between images of the future and societal change (Polak 1973), James Mau in his work on representations of future progress and political action (Mau 1968), Wendell Bell and James Mau in their collection of essays on the sociology of the future (Bell & Mau 1971), and Boulding in his work on societal images and the integration of the social sciences (Boulding 1966).

A second important perspective focuses on the social construction of time and the future. In this regard, the temporal order of social life has been studied (Zerubavel 1981), as well as the social processes that structure social times (Sue 2002). Other authors have explored the relation between a (weakened) social order which affects the individual capacity to influence and shape individual futures (Bauman 2002) and the social distribution of expectations (e.g. Brown et al. 2000 on socio-technical expectations). Also social theory has explored the ties between the idea of time and human cognitive agency (Zerubavel 2005) or the human capacity for mental synthesis and symbol formation (Elias 1986). Other examples could be provided as regards the primary feature of the sociological understanding of time and future, as well as of sociological theorising about time and the future: namely diversity. Acknowledging this diversity and in the awareness that a comprehensive account would be far beyond the scope of this article, I merely refer to Barbara Adam's extensive work on time (1990, 1995, 1998) as an outstanding example of a social-theoretical analysis of the multidimensional notion of social time.

The third aspect that I would mention concerns anticipation. From this point of view, the statement that futures research marks the beginning of a deliberate scientific effort to anticipate societal futures is historically inaccurate. In fact, since its very beginnings, sociology has aspired to an explanatory and predictive power enabling a more informed and rational social action. To quote the father of sociology, August Comte's well-known *motto* was "from science comes prevision; from prevision comes action" (*Savoir pour prevoir et prevoir pour pouvoir*). According to Comte, beneficial social action will become possible only when the laws of motion of human evolution have been established, and only when the basis for social order and civic concord has been identified. Furthermore, very early in its history, sociology adopted a critical stance on its presumed predictive power and engaged in a deep epistemic and methodological self-critique (today, we would probably define this stance as "reflexive"). To cite only a few examples, this perspective was assumed by Ludovico Limentani (1916) from a still positivistic perspective, and by Antonio Gramsci (1933/2001) from a Marxist and critical one.

As I have said, this short introduction lays no claim to comprehensiveness in framing the sociological perspectives on time and the future, on the one hand, and the interrelations and mutual influences between futures studies and sociology on the other. Rather, it selects some broad themes that seem relevant to describing how FS has acknowledged the centrality of the human and social dimension in exploration of futures, and how this has been inspired also by sociological thought.

Futures Studies (FS from now on), as a general term, will be used as defined by John McHale (1969) without specifying the differences between futures research, forecasting, foresight, prospective, etc., to be found in many texts (Barbieri Masini 1993).

The following sections will focus on the relations between social sciences and FS, on their human and social perspective, on some of their important aspects, on requirements for their development, and on their role in a changing world.

2 Social sciences, sociology and futures studies: the need for interdisciplinarity and possibly transdisciplinarity

Before I discuss the human and social perspective in FS, I must briefly state why it is important for social sciences, sociology and FS to acquire an interdisciplinary approach so that they can take up the challenge raised by interrelated and widespread changes, as well as to acquire a human and social perspective. The need for at least an interdisciplinary approach has been recognized by scholars in many sciences, as well as in social sciences and in sociology, mainly when they have looked at the future; and all more so in FS from a social and human perspective, where an ethical approach is important.

Interdisciplinary programs sometimes arise from sciences and new research developments. This is the recent case of nanotechnology, which cannot be addressed without combining the approaches of at least two or more disciplines.

This need among scientists is evident in various European Community projects, notably the HLEG (High Level Expert Group) project (Nortman 2004) which developed an interdisciplinary approach between the biological and hard sciences as well as the social and human sciences. It also developed a future oriented perspective.

Interdisciplinarity is the endeavor by two or more academic disciplines to work together in pursuit of common understanding of the complexity of an issue, whatever its area: the environment, education, health, work, the family, etc. The interdisciplinary approach is of great importance for the social sciences and sociology because the social environment is growing increasingly complex and its areas ever more closely interrelated. This social environment must be analysed using more than the knowledge and methodology of any single discipline.

Wallertstein (1996) wrote that trends towards the disciplinarization and professionalization of knowledge will not be sufficient in the social sciences and sociology because these seek to understand the changing social context and therefore require an interdisciplinary approach.

It is evident in this perspective that as sociology, and FS, seek to understand change, they must strive for interdisciplinarity. This challenges the various sciences to emerge from their niches, even at the risk of the insufficient scientific rigour that some social scientists have stressed. It is for this reason that FS have sometimes lost credibility with respect to other social sciences. The idea that FS needs social science, and especially sociology, is important because, first and foremost, the discipline seeks to understand social change from whatever point of view. This is not to say that the natural or hard sciences are unable to contribute to FS. Quite the contrary. Suffice it to recall a scientist like Dennis Gabor (1964) who wrote "Inventing the Future" and whose invention of the hologram gave great impetus not only to science but also to FS. Edgar Morin has subsequently used the image of the hologram to provide a highly imaginative description of the world situation that so many futurists have described in other ways. Morin writes that the hologram is the representation of a world where "not only every part of the world is more and more part of the world, but the world as a whole is more and more present in each of its parts" (Morin and Kern 1996, p. 22).

Transdisciplinarity extends beyond interdisciplinarity and is well known in social science (Masini 1983, p. 18 and Barbieri Masini in 2000 (2)). In fact FS needs not only the parallel approaches of different disciplines but also the joint effort of many approaches to address the complexity of current problems in a rapidly changing society. Transdisciplinarity entails that disciplinary approaches must find common assumptions and use common methodologies. Examples of such methodologies are Delphi and the scenario building methods that require the support of mathematics, statistics, as well as of sociology, economics, history and psychology. To quote Fred Polak: "all kinds of separate, fragmented portions of the jigsaw puzzles are of little avail, unless they are fitted together in the best possible way, to form an image of the future depicting a number of main areas of development" (Polak 1973, p. 261). Efforts towards a trans-disciplinary approach should be undertaken in the social sciences as well as in sociology and FS; but it is a long-term process, as all those working for it well know, whatever the area.

3 Why a human and social perspective for futures studies

On surveying future thinking in the past from a historical perspective, it is interesting that Plato, in his *De Republica*, already conceived the future of society as centred on justice, which would also be the core of social institutions. Saint Augustine's description of the future foresaw 'The City of God' as giving sense to human history. Another interesting example from the past is provided by Thomas More, the English philosopher, whose *Utopia* envisaged a society in which the common good was the central value. This latter view has been closely embraced by thinkers who look at the future from a social and human perspective. There are numerous historical examples in which looking into the future has meant considering the future of society and of human beings with the focus on ethical aspects, and hence on values. This topic is addressed in the next part of this essay.

After the Second World War especially, FS, or 'forecasting' as it was mostly called at the time, meant looking at the future from a scientific point of view, and using quantitative data as much as possible. This is understandable, for the War had demonstrated the need for forecasting technologies and armaments. This way of looking at futures was developed mainly in the USA, where, however, it very soon became clear that society must be the context of any forecasting. Among such thinkers I recall in particular John McHale and Alvin Toffler.

Also to be remembered is the extremely interesting discussion concerning the scientificity of forecasting developed initially in the 1960s by Olaf Helmer (1983). Helmer stressed that when we look at the future, we do not have quantitative data and we cannot use experimental, and therefore scientific, tools. This applies to all the social sciences, which cannot be strictly scientific because the choice of topic and indicators are intrinsically tied to a personal or social perspective and hence cannot be subjected to experiment as "a scientific approach" requires. Helmer wrote that we can at most speak of 'pseudo-experimentation', meaning experimentation based on models of reality. Another important point made by Helmer concerned the difference between the exact (physical sciences) and inexact sciences (human and social sciences). What the social sciences and FS can do is use tools and methods possessing scientific rigour in all their steps from clear statement of the issue under analysis to the chosen aim towards which future developments are directed. It is hence essential to frame the subject and the aim correctly, and to make rigorous use of methods.

In my view, therefore, it is important to accept that FS, sociology and social sciences have a component which relates to values, and hence to choices. This is why the specification of human and social FS is important. Within the framework of the approach just outlined, this essay will address the need for FS to encompass a philosophical and ethical perspective.

In this connection, the relations between sociology and FS indicated in the introduction of this essay must also be taken into account. Wendell Bell, a sociologist as well as a futurist, addresses this issue in his two volumes on the *Foundations of Futures Studies*. Specifically, in his chapter on "Is Futures studies an Art and a Science?" (Bell 1997), he stresses that "Futurists are interested in making social action more intelligent, informed, effective, and responsible. Although they may spin dreams of the future to help orientate that action, they are obliged to seek the truth. For intelligent, informed, effective, and responsible action requires reliable and valid description of present realities and reliable and valid knowledge of causes and effects" (Bell 1997, vol. I, p. 172). It is interesting to note that the title of the chapter recalls the theoretical debate in the 1960s between Bertrand de Jouvenel (1964) and Ossip Flechteim (1966) on futures studies ('futuribles' for the former, 'futurology' for the latter).

The above passage, in my view, shows that FS should conduct valid analysis of the present as well as possessing a good understanding of the past. They hence require a contribution from sociology and the social sciences at large. The passage by W. Bell also stresses the responsibility of futurists, thus highlighting a very important ethical component of their work, and at the same time, a need for value clarification as part of their responsibility in describing future possible developments.

In regard to the relations between sociology and FS, I would recall that, within the International Sociological Association (ISA), the Research Committee for Futures Research (RC07) which was founded in 1970 by Bertrand de Jouvenel, an economist but also an interdisciplinary social scientist and a futurist, has been, and still is, very active in its endeavour to develop dialogue and cooperation between sociologists and futurists (Barbieri Masini 2000 (1)).

In this context I wish to recall a phrase cited by the sociologist, Daniel Bell, a member and coordinator of the Commission Towards the Year 2000 established by the American Academy of Arts and Sciences in 1966. Bell quotes St Augustine as follows: "Time is a threefold present: the present as we experience it, the past as a present memory, and the future as a present expectation" (Bell 1968). It is interesting to note that members of the Commission included sociologists, like David Riesman and Wilbert Moore, political scientists like Karl W. Deutsch and Samuel Huntington, economists like Wassily Leontieff, and the anthropologist Margaret Mead. It was a truly interdisciplinary group comprising a large number of social scientists, and it can very usefully be revisited to see how an interdisciplinary group can work together, as well as to see how many of the issues that it foresaw and debated in 1967 are present still today. Notable is the case of the political scientist Fred Charles Iklé, who raised the issue in his article "Can Social Predictions be Evaluated?" (1968), stressing the need to take account of evaluation, as well as the distinction made by Robert Merton (1994), between self-fulfilling and self-defeating predictions and the consequent need to consider the effects of predictions. This is an issue that, in our age of ever more rapid information, brings to the fore the responsibility of futurists, especially in the social domain, and hence the need to take account of their value choices.

4 Futures studies and philosophical and ethical perspectives

In his seminal text "L'art de la Conjecture", Bertrand de Jouvenel (1964) clearly distinguished between what he called 'Facta' and 'Futura', the former being events that have already occurred (whence the importance of social sciences for analysis), the latter being events yet to occur. Although 'facta' are susceptible to values in the choices made by the researcher in terms of area variables, 'futura' have a higher presence of value choices.

The philosopher Peter Henrici has written that the object of human sciences, and of social sciences, depends partially on human choices, which are free, and hence may not be foreseeable. It is for this reason that FS should always look at the future in alternative terms, considering the possibles and probables as well as the desirables. Hence, Henrici writes, the future "is the consequence of free decisions which introduce themselves, into a predetermined texture of conditionings, changing it. In other words, the future must be conceived as a range of possible alternatives each of which is well determined in itself" (Henrici 1977, p. 28). On this basis, Henrici describes three approaches to the future: prognosis (the German term which I would define as forecasting), utopia, and project building.

The three approaches are centred on different levels of value presence: prognosis is based on the past and on analysis of the present, and hence on data, information and knowledge. This approach conducts analysis as strictly scientific as possible in the social sciences. Values are largely reflected, in this case, in the choice of areas, variables and indicators, and hence of quantitative as well as qualitative data (Barbieri Masini 1993, pp. 45-6). This approach entails looking at the future in terms of what is possible and what is probable. It can be said that future thinking in prognosis is 'retrospective' or 'a history of the future'. Prognosis in FS, being based on present and past analysis, in some way maintains "the same order."

Utopia is the second approach. It of course has a long history in the writings of such great philosophers as Plato, Thomas More and Francis Bacon, as well as the utopian scientists of the eighteenth century.

Utopia "is an *ideal model* of the world as it should be, while anti-utopia is the model of a world which should *not* be" (Henrici 1977, pp. 29-30). Utopia or anti-utopia are a-historical: "utopia is the building of a future different from the present. The data of the past and the present are not important, although if we analyse the different utopias, we see they are related to the present and are often in contradiction with the present, what is important is invention, innovation, imagination" (Barbieri Masini 1993, p. 45).

It is interesting to recall, in this regard, Ashis Nandi's argument in his book *Traditions, Tyranny and Utopias, Essays in the Politics of Awareness* (1987) that utopias are liable to be authoritarian if not open to criticism and self-criticism. In short, prognosis as well as utopia, are not sufficient in themselves to be effectively future builders.

The third approach relates to prognosis as well as to utopia, and builds a bridge between them. Utopia may clarify goals for the future, but it is not achievable unless prognosis gives a good analysis of the past and the present. Prognosis, in its turn, is only more of the same, past and present, with a low level of value presence and hence goals.

I call this third approach 'vision' while Henrici calls it project. It takes into account the past and the present, while at the same time searching in the present for what can realize the utopia. It comprises both prognosis and utopia, with the latter capturing trends from which the vision may receive strength and weakness The vision, on this account, is also a step towards the creation of a project because part of it is the will to realize the vision; thus it has within itself the capacity to build the future (Barbieri Masini 1993, p. 46). For a vision to become a project of the future, there must be a conscious will for it to become one. Here the responsibility of the futurist emerges, as well as of those interested in using futures studies.

At this point, the ethical component of futures studies appears with the presence of values, choices, and responsibilities at all three levels, but with different magnitudes.

The differences between the three approaches can be clarified as follows. In prognosis, the value component consists mainly in the choice of area and variables; in utopia, values are the basis of future thinking; in project, or vision, values and choices are present for the purpose of clarifying the goals. At the same time, the capacity of vision to depict the possible (or probable) futures is based on the best possible social analysis furnished by social sciences, and especially sociology. This is why, in my view, the connections between FS and social sciences and sociology are important.

At this point, the major task of futurists emerges as the search for what is not yet visible in the past and present but may constitute the 'seeds of change' or, as they are called by many, 'weak signals' which may in the future enable realization of the chosen goals, hinder them, or even destroy them.

I conclude this part with the phrase by the theologian Josef Fuchs which states that FS, as a discipline, is not only a possibility or a need, but a real and clear moral duty (Fuchs 1977, pp. 137-148).

5 The importance of images and visions in futures studies

The importance of the philosophical and ethical characteristics of FS, mainly in terms of a human and social perspective, has been described in the previous chapter, and so too has the connection, in these terms, with sociology and social sciences. I wish now, from this perspective, to expand on the role of vision, which I have described as the third approach of FS.

In this framework it is important to enumerate the differences between images and visions in FS. Images have been more debated in the history of FS than have visions, and both terms are used with different meanings. Because a specific view of visions is assumed in this essay, it may be useful to outline the difference between images and visions.

I start by recalling that Bertrand de Jouvenel (1964) stressed the importance of imagination in analysing possible, probable and desirable futures (futuribles). Robert Jungk (1969) and Dennis Gabor (1973) referred to "inventing the future." The scholar who has most profoundly discussed images is Fred Polak, who writes: "The formation of images of the future depends upon an awareness of the future that makes possible a conscious, voluntary, and responsible choice between alternatives This means that the development of images of the future and ethics are intimately related" (1973, p. 13). Elise Boulding, who abridged Polak's book, as well as translating it into English, writes in her introduction: "Very few people are aware of the intellectual debt that the future-oriented scholarship and activity owes to Fred Polak, who was the first, in the post-World War II period, to undertake the difficult conceptual work of clarifying the role of the image of the future in the social process and at the societal level." (Boulding in Polak 1973, p. VII).

In his book Polak writes that: "The future not only must be perceived: it also must be shaped" (1973, p. 13). He also says: "Once he (man) became conscious of creating the future, he became a participant in the process of creating this future" (1973, p. 6). This concept adds to what has been said on visions previously in this essay.

Wendell Bell, in his books "Foundations of Futures Studies" underlines that "Some futurists study the ways in which images of the future influence human behavior and how that behavior in turn contributes towards making the future" (Bell 1997, p. 82). This is very closely related to Polak's thesis.

Bell also writes: "One link between images of the future and present behaviour has to do with an individual's ability to balance present and future justification" (Bell 1997, p. 82).

At this point, I wish to tie these thoughts concerning visions with my own thinking over the years (Barbieri Masini 2002). In the book *Visions of Alternative Societies* (Masini 1983), which I edited within the "Goals, Processes and Indicators of Development" project and constituted the proceedings of a meeting held in Mexico City as part of a project sponsored by the United Nations University and coordinated by Johan Galtung, I, together with Bart van Steenbergen, attempted a first definition of visions in the introduction to the book. The meeting in Mexico City was attended by many leading scholars who have greatly contributed to FS, Elise Boulding, Roger Garaudy, Ivan Illich, Ashis Nandi, and Ian Miles. The definition we attempted in that now distant past was the following: "visions of the future are the stimulus to change the present" (Masini & van Steenbergen 1983, p. 3).

In the same definition we recalled Fred Polak's phrase: "from the antithesis between the present and the imagined, the future is born."

To develop these concepts further, I shall return to the philosophical and ethical bases of future thinking, with the support once more of the philosopher Peter Henrici (1977, pp. 19-20). According to Henrici, we must distinguish among three levels: 1) something is changing, 2) something may change, 3) something must change.

The first level, *something is changing*, is based on the analysis of the past and the present. In this phase we need the strong support of social sciences (history, sociology, etc).

The second level is where the work of FS is particularly important. The best way to describe it is as *something may change if.* Here of great support is the thinking of many futurists in terms of possibles and probables.

At this level the work of futurists is to go beyond the present, using in-depth social analysis to search for what may possibly and probably happen. In this connection, a normative approach is evident, meaning by this that the presence of values and choices is stronger than at the previous level.

Finally there is the third level, which implies that *something must change.* This is the core of visions. This aspect is laden with values as it involves a search for something that must be changed on the basis of choices. This is where the knowledge of the past and the present join the desirable and build action, namely a project, on a choice.

In this context the sociologist Rajni Kothari appropriately referred to: "a dilemma that faces the futurist. As a reformer and a romanticist, which every futurist must be, he is guided by a vision whose fundamental credo is how to leave the past behind and re-mould the present towards a different world. As a skeptic and a scientist, however, he knows that a total break with the past is both an impossible and a dangerous proposition and that all he can hope to have is a better world" (1974, p. 1).

In Polak's terms, the connection emerges between images, which are at the historical basis of cultures, and visions which build projects for the future as well as for cultures.

In this regard, Javier Medina, in his book *Vision compardida de futuro,* expands on what has been thus far discussed about visions: "the future and the present are tied by the vision. The vision lies in the perception of change in the present and hence is in itself the possibility of change, of building a future which is different from the present: visions are multiple, visions are social constructions, and finally visions have an anticipatory and strategic function (Medina Vasquez 2003, pp. 158-61).

6 Limits to futures studies

I have described the thinking of some sociologists and social scientists on the future, how a human and social perspective has emerged in FS, the latter's character of intedisciplinarity, the importance of a philosophical and ethical perspective, and finally the role of visions in their capacity to change the future and as a basis to build futures. I believe it is also useful, in this chapter, to present some of the limitations of FS as seen mainly from a human and social perspective.

The first limitation has been at the centre of important debate since the sociologist Robert Merton spoke of 'self-fulfilling' and 'self-defeating' prophecies in the 1940s, followed by many other studies in the area, amongst others by Richard Henshel. Henshel defined the concepts thus: "Self-fulfilling predictions and self-defeating (suicidal) predictions – collectively termed reflexive or self-altering predictions – occur whenever the issuance of a social or economic prediction causes alterations in behaviour (decisions) that promote or thwart the expected outcome" adding that there are "serious inherent limitations on scalar predictive precision in the social sciences" (1993, p. 2).

Henshel concludes that: "The first extensive treatments of the problem are those by Robert K. Merton in sociology and Oskar Morgenstein in economics. Right away the critical difference in perspective between the two fields appears in the work of these seminal writers. Merton emphasizes how self-alteration can increase predictive accuracy and Morgenstern emphasizes its possibly catastrophic diminution of accuracy" (1993, p. 2) Morgenstern's work was written in the 1920s.

A second limitation to futures studies resides in the psychological approach. Accepting that futures are alternatives is difficult for people in their daily lives, over and beyond thinking about the future itself. When people do think about the future in their daily lives, they tend either to be afraid of the worst future or naive about the best possible futures. As field research has shown, it is difficult for people to make the connection between the personal and the social in everyday life. Moreover, it is difficult to think of the future in alternative terms. This limit is often called the psychological limit.

To face this difficulty from a very early age, experiments have been carried out by educators, who are also futurists, to help children at school to think of desirable or feared futures in terms of 'what if'. Among them is Jane Page (2003, p. 2), who writes: "The issue of whether futures concerns might be beyond the reach of four and five year olds is central to any discussion concerning the formulation of a futures-based early childhood curriculum. Young children's attitudes towards the future and attendant notions of time and change differ fundamentally from those of adults. Yet their views should not be undervalued because they do not correspond to our own. They already possess many of the qualities stressed by adherents of futures studies as beneficial for a positive understanding of the future. Their flexibility of thought, their positive and constructive outlook on life, their sense of the continuity of time, their creativity and imagination and their sense of personal connection with time and the future, are all qualities which futures studies strives to re-instil in adults and older children. Their perspective on time and the future thus forms an ideal basis for a futures-focused curriculum" (Page 2000, p. 2).

Another psychological limit to future thinking is the fear of the unknown, which seems to increase with age as well as with changes in the economic, political and social environments in which people live. It is useful to recall at this point the phenomenon highlighted by Harold Linstone when he described "discounting the future" by people – and I would say also decision makers – who prefer not to consider the changes that may occur in relation to events or their own actions (1984).

Another important limitation of futures studies is the presence of 'implicit hypotheses'. This shortcoming is infrequently acknowledged because certain assumptions of future thinking may not be clear. For example, in the 1970s the assumption was that economic growth would continue as it had done since the end of the Second World War. As well known, this was the cause of some useless forecasts made at the time. A hidden hypothesis very often propounded in recent times is that technological changes will bring only benefits to humanity, without distinctions among different contexts. This is a further reason for the important ethical component that futurists should consider in their work. Hence once again important is the responsibility which concerns the need for value clarification by futurists as well as awareness of the consequences that research on, or indications of, possible futures may have on individuals and societies.

I conclude by saying that all futurists, and not only those supporting a human and social perspective, should be aware of these limitations to their work, as well as of their responsibility for the consequences of research and studies on society and on individuals.

7 Conclusion

To conclude the essay, I would say that the support of social sciences, and especially sociology, is very important for FS to build a sound basis for their reliability, as well as to depict trends emerging from the past and the present and which may develop in the future. Such support is important to perceive "seeds of change" and enable futurists to capture the possible and probable future, which may or may not be the one desired. This is where utopias and value choices are useful. As a connection between social analysis (prognosis) and utopia, the role of vision or visions emerges and may become a project or a multiplicity of projects. Visions are strongly related to value choices and to the desire for, or fear of, different futures by whoever builds them: individuals, decision makers, project leaders, social political or economic leaders. But they need social analysis as their basis. It should consequently not be forgotten that the changes desired or feared should arise on the basis of both social analysis that is as rigorous as possible, and of explicit goals supported by a sense of responsibility among futurists and decision makers. This is why the ethical component of FS is so important. On these bases it may be that environmental, social, cultural, economic or political changes are perceived, and visions and projects constructed, before serious damage has been caused to societies and people. It is hence important to repeat that FS are not only a discipline but also a moral duty.

References

Adam, B. 1990, *Time and Social Theory*, Polity Press, Cambridge.
Adam, B. 1995, *Timewatch. The social analysis of time*. Polity Press, Cambridge.
Adam, B. 1998, *Tim escapes of modernity. The environment and Invisible Hazards*, Routledge, London.
Barbieri Masini, E. 1993, *Why Futures Studies?*, Grey Seal Books, London.
Barbieri Masini, E. 1999, *Futures studies and Sociology, a Debate, a Critical Approach and a Hope"*, *International Review of Sociology*, 9 (3), pp. 325-331.
Barbieri Masini E. & Wilenius M. 1999, *"Futures Studies and Sociology: a Debate, a Critical Approach and a Hope"*, in: International Review of Sociology, Vol. 9, No. 3.
Barbieri Masini, E. 2000 (1), "Futures Studies and Sociological Analysis", *The International Handbook of Sociology*, ed. S. Quah & A. Sales, Sage, London.
Barbieri Masini, E. 2000 (2), "Future Studies as Human and Social Activity", Edgar F. Borgatta, Rhonda J. V. Montgomery (eds.), *Encyclopedia of Sociology*, Second Ed., Macmillan Reference USA, New York, vol. 2, pp.1037-1043.
Barbieri Masini, E. 2002, *Vision of Cultures, Vision Week -conference in Turku*, available at: http://www.tukkk.fi/tutu/Vision and http://www.tukkk.fi/tutu/Vision
Bauman, Z. 2002, *La società individualizzata*, Il Mulino, Bologna (*The Individualized Society*).
Bell, D. 1968, *Towards the Year 2000: Work in Progress*, Houghton Mifflin, Boston.
Bell, W. & Mau J.A. (eds) 1971, *The Sociology of the Future*, Russel Sage Foundation, New York.
Bell, W. 1997, *Foundations of Futures Studies*, 2 vols., Transaction Publishers, New Brunswick,
Boulding, K. 1966, *The Image. Knowledge and Life in Society*. University of Michigan Press, Ann Arbor.
Brown, N. 2003, "Hope Against Hype – Accountability in Bio Pasts, Presents and Futures", *Science Studies*, vol. 16 no. 2, pp. 3-21.
Brown, N., Rappert B., Webster A. (eds), 2000, *Contested Futures. A Sociology of Prospective Techno-science*. Ashgate, London.
De Jouvenel, B. 1964, *L'art de la conjecture*, Edition du Rocher, Monaco.
Durkheim, E. 2005/1912, *Le forme elementari della vita religiosa*, Meltemi, Rome. (*The Elementary Forms of Religious Life*).
Elias, N. 1986, *Saggio sul tempo*. Il Mulino Bologna. (*Essay on Time*).

Flechtheim, O. 1966, *History and Futurology*, Meisenheim, Verlag Anton Hain, Arn Glam.

Fuchs, J. 1977, "Morale come progettazione del futuro dell'uomo", (Ethics as Future Building Project), in P.C Beltrao (ed) *Pensare il futuro* (Thinking the Future), Edizioni Paoline, Rome.

Gramsci, A. 1933/2001. *Quaderni dal carcere. Quaderno 11*, Einaudi, Torino (*Prison Notebooks. Notebook 11*).

Helmer, O. 1983, *Looking Forward: a Guide to Future Research*, Sage, Beverly Hills and London.

Henrici, P. 1977, "La futurologia perché e come?", P.C. Beltrao (ed) *Pensare il futuro*, Edizioni Paoline, Rome *(Futurology Why and How?)*.

Henshel, R. L. 1993, "Do Self-Fulfilling Prophecies Improve or Degrade Predictive Accuracy How Sociology and Economics Can Disagree and Both Be Right", *The Journal of Socio-Economics*, Vol. 22, no. 21 (abstract).

Iklé, F. C. 1968, "Can Social Predictions be Evaluated?", D. Bell (ed), *Towards the Year 2000: Work in Progress*, Houghton Mifflin, Boston.

Kothari, R. 1974, *Footsteps into the Future*, Institute for World Order, New York.

Limentani, L. 1916, *La previsione dei fatti sociali*, Torino (*The Forecast of Social Facts*).

Linstone, H. 1984, *Multiple Perspectives for Decision Making*, North Holland, New York.

Masini, E. 1983, *Visions of Desirable Societies*, Pergamon Press, Oxford.

Mau, J.A. 1968, *Social Change and Images of the future*. Schenkman Publishing Company, Cambridge, Ma.

Medina Vasquez, J. 2003, *Vision compartida de futuro*, Universidad del Valle, Cali *(Shared Vision of the Future)*.

Merton, R. 1940, Bureaucratic Structure and Personality, *Social Forces XVII.*

Nandi, A. 1987, Traditions, Tyranny and Utopias, Essays in the Politics of Awareness, Oxford University Press, Oxford.

Nordmann, A. 2004, Converging technologies. Shaping the future of European Societies, European Commission, Bruxelles.

Page, J. 2000, Reframing the early Childhood Curriculum: Educational Imperatives for the Future, Routledge, Falmer, London.

Polak, F. 1973, and abridged by Elise Boulding, The Image of the Future, Elsevier, Amsterdam, London and N.Y.

Sue, R. 2001, *Il tempo in frantumi: sociologia dei tempi sociali*. Dedalo, Bari. (*The Broken Time: a Sociology of Social Times*).

Wallerstein, I. 1996, *Open the Social Sciences*, Standford University Press, Standford.

Weber, M. 1991/1905, *L'etica protestante e lo spirito del capitalismo*, Rizzoli, Milano. (*The Protestan Ethic and the Spirit of Capitalism*).
Zerubavel, E. 1981, *Tempi nascosti: orari e calendari della vita sociale*. Il Mulino, Bologna (*Hidden Rythms. Schedules and Calendars in Social Life*).
Zerubavel, E. 2000, *Mappe del tempo. Memoria collettiva e costruzione sociale del passato*, Il Mulino, Bologna (*Time Maps: Collective Memory and the Social Shape of the Past)*.

4 Dealing with the future in economics

Cornelius Hazeu

1 Introduction

This article addresses the question of how economists deal with 'the future'. In answering that question, I shall focus chiefly on the macroeconomic preparation of government policy aimed at the future, and on the role of economists and economic institutions in that process. After this brief introduction, Section 2 looks at the specifics of the economic discipline and at a number of characteristics of the economic method that are relevant for the subject of this article, namely how economists deal with the future. Section 3 discusses the developments that have taken place since 1945 in particular, and which have led among other things to the creation of specific institutions which explore the economic future. Section 4 considers a number of methodological problems, in particular the 'Lucas Critique' of the use of models in preparing economic policy. I then take a look at how contemporary modelling of the economic future deals with institutions (section 5). The article ends with a number of conclusions (section 6).

History box: some famous economists about the future

Adam Smith (1723-1790)
Never people will live in Utopia. Human characteristics, behaviour and contradictory interests exclude that perspective. Therefore it's no use designing an ideal society. But according to Smith there is no need to be pessimistic all the way: the working of markets, combined with governments that deliver safety and security in society and a limited number of other government tasks, can stimulate welfare growth and prosperity.

David Ricardo (1723-1790)
Ricardo's work was concentrated on the different production factors, their ideal prices, and the allocation of production over time. Ricardo appears to regard public expenditure as unproductive and adopts Say's maxim "that the very best of all plans of finance is to spend little, and the best of all taxes is that which is the least in amount". Also government expenditure for public works – aimed to yield future benefits – cannot do much good: "the raising of funds for the purpose of employing the poor......diverts those funds from other employments which would be equally if not more productive to community".

Joseph Schumpeter (1883-1950)
According to Schumpeter ups and downs in economic development can be explained by the fact that new combinations or innovations appear. The *entrepreneur* who applies new combinations of factors of production, plays a central role. He is the innovator, and the agent of economic change and development. Centred around the role of the Schumpeterian entrepreneur, is the rise and decay of capitalism. Schumpeter predicted the gradual decay of capitalism. In the contary with his contemporary Keynes, Schumpeters ideas were concentrated on the long term cycles and developments in economic systems.

John Maynard Keynes (1883-1946)
"In the long run we're all dead."

Milton Friedman (1912-2006)
Only by the correspondence of the predictions and the facts should theories be provisionally accepted or rejected. Results, not assumptions should be the main focus of our scientific activity in understanding the real world.

2 Modelling the future means using 'if-then' reasoning as the core of economists' methodology

Economists are constantly concerned with mapping out future trends: what will the rate of economic growth be next year? What will population ageing mean for the public finances? How big will the market be for a new product? And so on. The *future* might be by definition unknowable, but this does not take away the constant need of governments, businesses, citizens/consumers to seek guidelines with which to underpin the decisions that they have to take *today*. Many of those decisions are financial/economic in nature, whether we are considering the purchase of a new car or a new commercial building, or equally where we have to decide whether or not to embark on a course of study, for example, or, conversely, to end a course of study once begun. It is for this reason that economists make such efforts to try and foresee the future, for example trying to calculate when a new commercial building will become profitable, or what your future income could be if you follow a certain course of study, or what opportunities you will miss if you break off or choose not to follow a particular course of study, etc. Questions such as these are the core business of economists. In order to provide substantiated answers to them, special institutions have even been created, for example in the form of government planning offices, or strategic units within companies.

It cannot be assumed from the fact that economists are so completely occupied with developments in the future that they are also able to *predict* that future (accurately). To the extent that they venture to do so, they restrict themselves to estimates of economic key figures – interest rates, economic growth rate, etc. – and prefer to forecast those figures for the next quarter or the next year rather than for the year 2010, for example; because economists (like everyone else) are then no longer on firm ground. Over the long(er) term, too many things can change simultaneously, whereas an economist must be able to start from the basis of a number of fixed values in order to be able to extrapolate variables within that framework.

How economists deal with the future is in itself a very wide-ranging theme, because there is a very great deal of economy. In this article, approaches from the perspectives of business economics and management theories are left out of consideration (for more on these aspects see the article by Den Hartigh and Van der Duin elsewhere in this book) and concentrate instead on general economics. When general economists want to say something substantiated about future trends, they tend to use macroeconomic models: "economists do it with a model". In those models, observed correlations between economic variables are systematised in the form of (a large or small set of) equations. Precisely because those correlations are based on a large number of conditions and assumptions, economists prefer to use terms such as 'exploration', 'exploratory study', etc., for their activities focusing on the future – we prefer to avoid terms such as 'prediction'; it has too many connotations of 'crystal-ball gazing' and does not properly reflect the uncertainties inherent in explorations of the future. Economists are thus concerned with *exploring* a possible trend – or several possible trends if a scenario-based method is used – on the basis of assumptions about correlations observed in the past. The term 'no-surprise' exploration is also sometimes used for this, to indicate that no allowance has been made for unforeseeable exogenous events (major global environmental disasters, wars, etc.).

This method of future exploration using macroeconomic models assumes rationality, repeatability and predictability in the actions and behaviour of economic actors. This is typically something that under normal circumstances perfectly well *can* be assumed at macro-level because of the large numbers involved. However, these plausible macro-assumptions do not alter the fact that at the underlying micro-level, rationality is by no means always necessarily the rule. This is the familiar *fallacy of composition*: the whole can actually be less than the sum of its parts. When it comes to human behaviour at micro-level, we all know from our own experience and/or from those around is that people sometimes simply do something without thinking it through properly, or deliberately do something differently from what they did on a previous occasion in a similar situation, or do not look beyond the end of their noses, sell themselves short, do not learn or do not react to (new) information. All these forms of *myopic* behaviour which work against the rationality assumption of economists may be present at the micro-level, but this does *not* alter the fact that if all micro-level behaviour is aggregated, recognisable patterns do emerge at macro-level which generally hold water. For example, if we constantly see that in the past income growth of 100 has been accompanied by a growth in consumption of 90, then this will also be the most likely outcome in the future. In any event, if there is no special reason to deviate from a correlation observed in the past, this is then the most plausible assumption. Economic models are full of these kinds of empirically validated assumed correlations.

Apart from this, there is no reason to narrow or restrict the economic notion of rationality to people's reactions to financial incentives; Titmuss (1970) and Le Grand (2003), for example, have shown that people's motivation structures are much broader than this, and that there are domains in which financial incentives play no role, or even play an adverse role (Wolfson 1988, pp. 110 ff.; Hazeu 2007, pp. 91 ff). In the upper echelons of the economic and political system, attention for this aspect has to date been (sadly too) limited.

What all this boils down to is that the economic method of approximating future trends is characterised by *if-then* reasoning (Robbins 1935). This does not mean that we *know* the future, but it does enable us to make best guesses. This is expressed by adding a measure to an economic equation which represents the strength of the correlation found; this is the R2 factor, which represents the correlation and therefore gives an indication of the probability that an observed correlation will occur again in the future (cf. Morgan, 1990, pp. 28-30).

It is clear from the foregoing that statements by economists about the future are never absolute, but are always *conditional on the assumptions*: it is always as broad as it is long. The fact that economic models are based on assumptions rather than certainties, and that those assumptions are made explicit, is not a weakness, but is actually their strength. It is a way of rendering account; it makes model-based arguments explicit and verifiable for third parties. It is the counterpart in economic models to the 'if-then' reasoning.

3 From plans and predictions to explorations

There is no other period in history when the idea of a plannable and makeable society was as strong as it was in the first quarter of a century immediately following the Second World War (1945-1970). The welfare state blossomed in the West in that period. Governments assumed concrete economic tasks and objectives in relation to unemployment, inflation, the balance of payments, etc. And because national economies at that time were still far from being integrated, either globally or at European level, national governments were *able* to protect and screen off their economies to a large extent, and to pursue their own national macroeconomic policy. This relatively short – set against the background of history – era of potent national macroeconomic policy fostered a widely shared belief in the second half of the twentieth century in the idea of a 'makeable' society and a 'makeable' economy. Naturally, there were fluctuations in the economic cycle, but it was precisely for this reason that the Keynesian approach had developed and been refined: to smooth off and control those cyclical fluctuations. It was also against this background that the need for medium and long-range economic explorations arose, to answer the question what developments will take place in the future. If we have a better idea about that, we better know which buttons to press.

This also hallmarks the relation between politicians and economists: the politicians have their different ideological aims and priorities, with its consequences for the public finances; the economists make their calculations and show the limits and boundaries.

Starting from the idea that economic development was controllable in the short term, the policy – and the institutes involved in preparing that policy – began taking the next steps. To illustrate this, I will use the Netherlands as an example of a development which occurred in all Western countries, with at most slight variation between the institutions and a few national idiosyncrasies. The Netherlands Bureau for Economic Policy Analysis (CPB) was created in 1945, with the later Nobel Prize winner for economics, Jan Tinbergen, as its first director (1945-1955). The CPB became the leading economic planning and forecasting agency in the Netherlands, with the aim of helping the government to steer economic developments. In practice, the founders of the CPB quickly realised that planned development of the economy was an impossible dream, and the CPB's original 'design' ambitions were in a few years abandoned (cf. Passenier 1994). What remained was a focus on providing a scientific underpinning for policies, at least to the best of the CPB's ability.

Based on this premise, for developments in the short term the CPB publishes an annual Central Economic Plan *(CEP)* and a *Macro Economic Outlook (MEV)*. These publications describe the (small) expected movements in variables in the near term. The 'major' developments, by contrast, are depicted more clearly in the long-range studies published by the CPB. These studies gradually became more and more important, and some of them attracted a great deal of attention. In 1992, for example, the CPB published the study *Scanning the Future* (CPB 1992), which explored international economic developments for the period 1990-2015. The scenarios from this study were then translated into the implications for the Netherlands in *Nederland in drievoud* ('The Netherlands in triplicate', CPB 1992a).

These long-range studies employed several scenarios, each of which in turn had to be sufficiently plausible, and were published against the background of a broader trend in future research which has emerged strongly over the last thirty years. Alongside the CPB, the Scientific Council for Government Policy (WRR) was founded in the Netherlands in 1973, with the brief among other things of carrying out coherent long-range studies. These studies go further than the economic aspects alone, and in fact should ideally be so broad that they incorporate all aspects that are relevant for policy. For the *integral* long-range study for the year 2000, a task which the WRR assumed immediately after its foundation, the CPB was asked to carry out an economic preliminary study. This resulted in one of the first scenario studies by the CPB: '*Twee* perspectivische tekeningen van de Nederlandse economie tot 2000' ('*Two* perspective sketches of the Dutch economy up to the year 2000') (for a discussion of the quality of this study, a 'post-mortem analysis', see Den Butter 2004). The development of *two* future scenarios was strongly influenced by the Report for the Club of Rome which had been published some years before (Meadows 1972), with its disquieting message about world overpopulation, environmental pollution and exhaustion of natural resources. The impact of this report in the Netherlands, in particular – at that time the most densely populated country in the world – was great. That impact was also reflected in the CPB study for the WRR, in which a 'business-as-usual' scenario was adopted projecting a continued annual productivity increase of three percent, as well as a scenario in which stricter environmental standards reduced the growth in productivity to zero.

The principle of working with (several) scenarios was given a major boost at that time by the Long Range Planning Group at Shell (Schwartz 1991). The essence of this scenario approach is that the scenarios each describe a different potential future, based on variations in the development of critical uncertain variables. All scenarios must in principle be plausible, otherwise it becomes a put-up job. The idea is *not* that one of the scenarios could rapidly become a reality, because by definition the alternative scenarios in that case could not simultaneously be plausible. On the contrary, the intention of the scenario approach is to initiate a thought process, so that an organisation (in this case Shell) or a country is better prepared for new risks and new opportunities in its (market) setting.
The usual method used to create different scenarios is to develop a quadrant whose axes are formed by two sets of critical and uncertain variables (*driving forces*). This is also the method followed by the CPB in 1992 in *Scanning the Future*, in which – because the Global Shift and Global Crisis scenarios did not differ much in their implications for the Netherlands – three economic future scenarios were ultimately developed: *Global Shift, Balanced Growth* and *European Renaissance*. It is always interesting to carry out a post-mortem study in such cases: which of the – all equally plausible – scenarios turned out to be (most) true, and why?
More than ten years later the CPB published its next major scenario-based future study, in which the central focus was on the development of Europe as a steering entity (De Mooij & Tang 2003). The study was mainly concerned with the question of the possible responses to population ageing and the implicit increasing pressure on the public sector. The methodology differed slightly from the earlier study: the axis system was no longer made up of exogenous uncertain variables, but this time consisted of the different types of possible policy response: more or less international cooperation on one axis, and a spectrum from private to public cooperation on the other. This gave rise to four scenarios, within which the actions of policymakers were analysed. The policies actually pursued thus determine which scenario comes closest to reality. By introducing this endogeneity, the CPB did not make things easy for itself. The impact of this major future study was much less marked than it had been at the beginning of the 1990s, mainly because it was less easy to replicate it for the many other sectors in which the uncertain future is an issue. It was precisely this aspect that had in part given *Scanning the Future* its success (cf. Schoonenboom 2003).

4 The long term in economic models and the 'Lucas Critique'

As we saw earlier, economic models can enable pronouncements to be made about which – expected or possible or plausible – developments could take place in the short (within two years or less), medium (within three to five years) or long term (for example ten years or longer). Models can extend to cover the entire economy of a country or can be focused more narrowly, for example on specific sectors or regions. Economists do not use a single model to serve these different aims; 'different models for different purposes' is the credo. The model is after all not a crystal ball in which the future can simply be read, but is designed on the basis of the specific issues and questions with which it is concerned. Depending on the specific aims and context, an economic model is used to try and map out relevant correlations as accurately as possible. As stated, models are also always based on assumptions, which in turn are based on deeper-lying *paradigms*, such as the idea that the economy always tends towards equilibrium in the long term, or conversely, the idea that a permanent imbalance of markets is also possible.
The post-war macroeconomic model structure has led to ever better models in terms of the adequacy of their predictions. The builders and users of the models – often trained in the tradition of Keynes and Tinbergen – do have to have gone through the process of discovering that macroeconomics is not just a matter of 'push a button and watch it go'. That process is begun with attention for the fact that there are all manner of delays inherent in the economy, which reduce the visibility of policy input on the policy outcome. Delays occur in the decision-making process, and may for example mean that a tax cut designed to stimulate spending in the economy actually takes effect only when the economy has begun to improve again spontaneously. There may also be delays in the reaction of economic actors to policy changes, for example the consumer who does not immediately convert a tax break (or, for example, the release of monies accumulated in a salary savings scheme, to use a Dutch example from 2005) into extra consumption because he does not entirely trust the government.

This understanding that the psychology of those who are the targets of policy plays an important role is further crystallised in the doctrine of 'rational expectations': the actors in the economy see through the policy 'manipulations' of the government and go in pursuit of their own expectations of the future. In doing so they are rational enough to realise that a tax gift from the government now will have to be 'paid back' in the near future in order to compensate for the resulting budget deficit; hence the term 'rational expectations'. The

'Lucas Critique', named after the American economist Robert Lucas (1980), is grounded on this principle.
The reason for and background to the Lucas Critique was the fact that the early 1970s had ushered in a period in which it became increasingly clear that the economic policies of governments, based on the Keynesian models that had become the 'standard' in the years after the Second World War, were not having the envisaged results: (too) little effect, or too much, or too late. As a result it became the period, as could be judged later, which marked the beginning of the end of the Keynesian policy of controlling the economic cycle through tax and spend policies. The Lucas Critique played an important role here, especially in focusing the debate on the prevailing economic methodological assumptions. The actual transition from Keynesian economic policy to structural policy ('supply side economics') followed in the early 1980s: the time of Reagan (USA), Thatcher (UK) and Lubbers (the Netherlands).
When we try to foresee the economic future, one way of doing so is through a simple extrapolation of trend lines. This is the *naive prediction method*: yesterday the temperature was 20 degrees Celsius; let us therefore assume that it will also be 20 degrees tomorrow; over the last 30 years labour productivity has grown by an average of 2 percent per annum; let us therefore assume that this rate will continue in the coming years; and so on. Predictions can also be based on a statistical forecasting model with hundreds of equations; as stated earlier, in the Netherlands this is mainly the province of the Netherlands Bureau for Economic Policy Analysis (CPB). Whether we use simple or more complex models, however, in both cases *the basic assumption is that the structures that applied in the past will continue to apply in the future*. And that can be a shaky assumption, especially when it is applied to the more distant future. Often those structures will change over time: 'distorted' relationships arise, 'thresholds' are exceeded or 'ceilings' are reached, causing relationships to change. Sometimes these changes in economic relationships observed earlier form an autonomous process, but change can also happen under the influence of (government) policy which is based precisely on the model extrapolation. That extrapolation has then acquired a *self-denying* character; that is the heart of the Lucas Critique. Its practical implication is that the impact of a stimulus to the economy from the government (for instance tax cut, spending programme or a form of (de)regulation) will be much weaker than predicted by the model, chiefly because citizens and businesses have already anticipated it. The effectiveness of macroeconomic government policy, especially economic measures, can therefore be disappointing because the public and businesses are thinking: "Yes, taxes may be going down, but that's temporary; they'll

soon be going up again, so we are not going to consume (consumers) or invest (businesses) because of that tax reduction".
The Lucas Critique, also coined the 'policy invariance argument', has caused a lot of commotion among economists and elicited many reactions. It would be going too far to examine them all here. One result in any event appears to be that the Lucas Critique is much more relevant for the (analysis of) financial markets than for the labour market. In addition, the Lucas Critique has consequences for both policymakers and model-builders. The consequence for macroeconomic government policy is that it is less effective than policymakers might have assumed in advance. The consequence for model-building is that 'large' policy changes cannot be extrapolated within the structure of existing models, because the behavioural parameters change.

5 The role of institutions

A second important methodological point is that economists have gradually become more aware that institutions should be given a place in their models, and how this should be done, even if they are only included as a delaying factor. Illustrative of this growing interest in the institutional setting in which economic actions take place is for example the fact that the Netherlands Bureau for Economic Policy Analysis (CPB) has since the early 1990s been intensifying part of its future research on the Dutch economy in this direction; this is apparent from the *Scanning the Future* exercise as well as from later future studies (De Mooij & Tang 2003). In the 'Germany study' by the CPB for instance (CPB 1997), institutions are actually the central focus. The 'Germany study' is concerned with the institutional differences between the Netherlands and Germany across the breadth of the labour market. This 'parallel study' of comparable (nabour) countries generates recommendations for both countries with regard to institutions: which coordination mechanisms need to be improved, and how, in order to improve economic performance? Which institutions hold back economic performance due to high transaction costs?
In the recent macroeconomic theory and its application, attempts are increasingly made to take account of the importance of institutions. Institutions can reduce the transaction costs in the economy. This implies that, although they appear to be exogenous, in the long term their rise and fall is also influenced by (broadly defined) costs and benefits. Recognising the importance of institutions also implies that the assumed operation of certain instruments in the economy can turn out differently from in the imaginary 'institution-less' world.

Institutions can also get obsolete, for example because the outside world – the context – changes or the scale or structure of the problem alters. This causes the overall costs and benefits, in transaction terms, to change as well. For the builders of economic models the task is then to show what it 'costs' if such an institution is maintained. In the 1990s research was carried out in this light in the Netherlands on issues such as maintaining the minimum wage at family level, declaring collective labour agreements to be 'generally binding', or the continued existence of the marketing and commodity boards. It is not relevant for the argument propounded here to look in detail at the outcome of these and similar studies; what is important is that institutions have been 'discovered' in the mainstream of general economic (future) research (Hazeu 2007). This theoretical development also stands for a long term perspective on economic phenomena.

6 Conclusions

Economists do it with a model. That is how economists deal with the future. Models are based on assumptions and 'if-then' reasoning. That makes them verifiable, controllable and scientifically sound. What economists are striving for when they concern themselves with future developments, therefore, is to make pronouncements about the future that are sufficiently plausible. But probability is not certainty. There is a great deal that we do not know, but what we do know can be placed within certain margins of certainty.
There are also any number of conceptual pitfalls when it comes to incorporating the future in economic models. The outcomes of models designed to predict the immediate future are the most reliable. The most difficult thing, including for economists, is the real long term. The longer the period covered by the prediction, the less firmly the structural characteristics of the underlying modelled reality can be posited. This is where the 'Lucas Critique' comes in, which I discussed in section 4. The contribution of John Maynard Keynes lays in making the economic cycle, which normally lasts six to seven years, controllable through policy. Starting from that preoccupation, he had his own way of dealing with the distant future, as borne out by his famous statement: "In the long run we're all dead". In short, economists are actively engaged with the future and future developments, but *based on the economic method* (section 2); as a result, they have little in common with futurists. Or, to end with the words of Ludwig Wittgenstein: "Wovon man nicht sprechen kann, darüber muss man schweigen" ("What we cannot speak about, thereof one must be silent").

References

Butter, F.A.G. den 2004, 'Twee Perspectivische Tekeningen van de Nederlandse Economie tot 2000', in *Vijfentwintig jaar Later. De Toekomstverkenning van de WRR uit 1977 als Leerproces*, ed. P.A. van der Duin, C.A. Hazeu, P. Rademaker & I.J. Schoonenboom, WRR Verkenning nr. 5, Amsterdam University Press, Amsterdam.

Centraal Planbureau (CPB) 1992, *Scanning the Future*, Sdu Uitgevers, The Hague.

Centraal Planbureau (CPB) 1992a, *Nederland in Drievoud*, Sdu Uitgevers, The Hague.

Centraal Planbureau (CPB) 1997, *Challenging Neighbours. Rethinking German and Dutch Economic Institutions*, Springer, Berlin.

Duin, P.A. van der, C.A. Hazeu, P. Rademaker & I.J. Schoonenboom (eds.) 2006, 'The Future Revisited. Looking Back at The Next 25 Years by the Netherlands Scientific Council for Government Policy (WRR)', *Futures*, June 2006, vol. 38, no. 5, pp. 235-246.

Grand, J. le 2003, *Motivation, Agency and Public Policy*, University Press, Oxford.

Hazeu, C.A. 2007, *Institutionele Economie. Een Optiek op Organisatie- en Sturingsvraagstukken*, second, revised edition, Coutinho, Bussum.

Hazeu, C.A. 2007a, Het Toekomstconcept in de Economie, *ESB*, vol. 92, 24 augustus 2007, pp. 491-493.

Lucas, R.E. 1980, 'Methods and Problems in Business Cycle Theory', *Journal of Money, Credit and Banking*, vol. 12, pp. 696-715.

Meadows, D. 1972, *The Limits to Growth. A Report for the Club of Rome Project*, Universe Books, New York.

Mooij, R.A. de, & P.G.J. Tang 2003, *Four Futures of Europe*, Koninklijke De Swart, The Hague.

Morgan, M. 1990, *The History of Econometric Ideas*, Cambridge University Press, Cambridge.

Passenier, J. 1994, *Van Planning naar Scanning; een Halve Eeuw Planbureau in Nederland*, Wolters-Noordhoff, Groningen.

Robbins, L. 1935, *An Essay on the Nature and Significance of Economic Science*, second edition, MacMillan, London.

Schoonenboom, I.J. 2003, 'Toekomstscenario's en Beleid', *B&M. Tijdschrift voor Beleid, Politiek en Maatschappij*, vol. 30, no. 4, pp. 212-229.

Schwartz, P. 1991, *The Art of the Long Term View*, Century Business, London.

Titmuss, R. 1970, *The Gift Relationship*, George Allen and Unwin, London.

Wolfson, D.J. 1988, *Publieke Sector en Economische Orde*, Wolters-Noordhoff, Groningen.

5 On the philosophical foundations of futures research

Joseph Voros

1 Introduction

This chapter considers the philosophical foundations upon which futures research is undertaken. It does not consider the views that different philosophies take of the future – that could easily be the subject of an entire series of books. Rather, the interest here is the way that different philosophical positions lead to different ways of undertaking knowledge inquiry, including *inter alia* futures research.

Futurist Willis Harman (1976) succinctly pointed out the central rationale of futures research:

> *[O]ur view of the future shapes the kind of decisions we make in the present. ... Every action involves some view about the future – as we expect it to be, or as we desire it to be, or as we fear it may be. If our image of the future were different, the decision of today would be different. (p. 1) ... Every action decision involves some assumption about the future; it is the function of futures research to make those assumptions explicit. Since we cannot know the future precisely, we must delineate alternative possibilities so that choices can be tested against various future states that could occur. But which futures are feasible and which are not? That is the central question of futures research (p. 10).*

This brief passage from one of the founders of modern futures research highlights the central role that ideas about, or 'images' of, the future have in all purposeful future-oriented human activity: these images of the future influence the decisions and actions we make in the present, which in turn have consequences in and for the future which eventuates as reality. There is a feedback loop between present ideas about the future, decisions made in the present, and the ultimate future which eventually emerges. However, this loop is not entirely closed – ideas about the future, and our decisions based upon them, do not always or entirely condition the future that eventually becomes reality. In other words, the future is *not* pre-determined (cf., e.g., Amara 1981). This indeterminacy with regard to the future is precisely what stops us being merely passive recipients of an unchangeable future, and gives us the latitude to become active agents who are able to shape the way the future unfolds, at least in part.

But how does futures research go about assessing alternative possible futures and generating futures-relevant knowledge? This is a complex question; many of the other chapters in this volume have also addressed it, in their own way, or through their own thematic lens. The purpose of this chapter is to consider the process of knowledge inquiry itself – it sketches some of the broad outlines and discusses some of the philosophical and methodological foundations upon which different approaches to knowledge inquiry rest, and briefly considers how these have historically influenced futures research.

It is considered axiomatic here that knowledge inquiry cannot be properly or even competently undertaken in the absence of an understanding of the philosophical foundations underpinning it. This is simply because any approach to generating knowledge is built upon an assumption base, which latter flows from certain foundational philosophical presuppositions – about the nature of reality, about the nature of knowledge, and so on. Researchers and practitioners need to be fully aware of this assumption base, in order to assess whether it is appropriate to and commensurate with the form, domain and purpose of the inquiry being undertaken.

To this end, a well-known typology of research or inquiry approaches is introduced and outlined as a basis for discussion. While a large number of approaches to inquiry exist, it is possible to conceive of these approaches as belonging to a few broadly-defined classes or categories (rather like the way that the millions of colours in the spectrum of visible light can be considered to inhabit seven or so main 'bands'). The typology can be considered to be a set of broad-brush-stroke generalisations which look for the overall large-scale structure of the wider landscape of inquiry approaches in general, while at the same time recognising that many gradations and inter-leavings exist between the various forms. From this understanding, we then briefly consider how the typology as defined can be seen reflected in the various forms of and approaches to futures work which have developed over the last few decades, ever since futures research emerged as a distinct field of endeavour in the middle of the twentieth century. This will require us initially to consider the nature and purpose of futures research, as well as the object domain with which it deals and, before that, to consider the role that philosophical insight has in the process of knowledge inquiry.

2 Why bother with philosophy?

Any approach to knowledge inquiry rests upon certain foundational assumptions and fundamental presuppositions – about the nature of reality; about the nature of the form of knowledge possible about that reality; about the types of methods which can be used to generate that knowledge; about the purpose of carrying out the inquiry; and several others. In other words, any formalised approach to knowledge inquiry (or 'paradigm', to use Kuhn's famous and often-misused term) engenders certain commitments and assumptions which are inherent in and constitutive of the paradigm – including ontological, epistemological, and methodological. Different paradigmatic foundational assumptions give rise to different forms of and approaches to inquiry, and these assumptions condition what are considered to be acceptable, appropriate or valid forms of methodology. Therefore, it is impossible to separate methodological considerations from the associated underpinning philosophical foundations, and to attempt to do so is, to quote Donald Michael (1985), to have "both feet planted firmly in mid-air".

It is useful here to recall an observation of Einstein's to a younger colleague concerning the need for philosophical insight in the scientific enterprise (Einstein 1944; Howard 2004):

> *I fully agree with you about the significance and educational value of methodology as well as history and philosophy of science. So many people today – and even professional scientists – seem to me like someone who has seen thousands of trees but has never seen a forest. A knowledge of the historic and philosophical background gives that kind of independence from prejudices of his generation from which most scientists are suffering.* ***This independence created by philosophical insight is – in my opinion – the mark of distinction between a mere artisan or specialist and a real seeker after truth*** *(emphasis added).*

In Einstein's view, therefore, philosophical insight – a deep understanding of the philosophical background and underpinnings of the search for truth and knowledge – is essential to avoid becoming merely an artisan or specialist – i.e., someone who is able to perform skilful actions, but without any understanding or ability to see beyond the prejudices of training, technique and historical epoch. This view will be taken as axiomatic here.

The contention here, therefore, is that methodological interventions cannot be properly or even competently undertaken in the absence of a solid understanding of the philosophical foundations underpinning them. This is because every methodological approach is founded upon a more-or-less implicit philosophical basis. It is therefore necessary for inquirers, researchers and practitioners to be fully aware of just what this basis is and whether it is appropriate to the form, domain and purpose of the inquiry. This is true for *any* researchers, and it is equally true for futures researchers. For, without such an understanding, we may easily become mere actor-artisans mechanically performing methodological activities that generate data, absent any real understanding of what those data are, mean, or how they may be contextualised in any larger framework of knowledge. Therefore, we must explicitly and consciously consider the philosophical foundations upon which futures research may be built. And to do that we must consider both the object domain as well as the objective of futures research: 'the future', and informed action.

3 Futures research as an 'action science' using 'images of the future'

Although a recognisable 'futures field' has existed since the 1960s (see, e.g., Amara 1974, 1981; Bell 2005; Linstone & Simmonds 1977), there is still some contention and debate over whether or not futures research can properly be called a 'discipline'. The push to do so may perhaps be fuelled by a desire to give futures research some sort of 'home' within the still-compartmentalised and largely discipline-based structures of modern schools and universities, and in the modes of thinking which are engendered by education based on this model of organising knowledge. There are influential and respected futurists on different sides of this 'discpline' debate (e.g., Bell 2002a, 2002b; Marien 2002a, 2002b). Michel Godet (2000) has even suggested that futures work is an "intellectual *un*discipline". In any event, one wonders whether a field which attempts to be consciously multi-disciplinary will ever find a comfortable home within a discipline-based model of academia. We can to some degree bypass this so-far fruitless debate by instead asking what futures research *does*, rather than asking whether it is a 'discipline' in the sense of a branch of knowledge with a defined object domain.

Futures research, by its very nature, is not and cannot ever be an empirical undertaking in the literal sense of the terms 'futures research' or 'futures studies'. As James Dator (2005) reminds us in his First Law of the Future:

> *'the future' cannot be 'studied' because 'the future' does not exist. Futures studies does not – or should not – pretend to study the future. It studies ideas about the future (what I usually call 'images of the future').*

Thus, it is not the (non-evidential and non-existent) *future* which is the object of inquiry in futures research – rather, it is the plurality of ideas about or images of the future which human beings have *in the present* which constitutes the object domain of futures inquiry. This is a subtle and frequently-overlooked distinction. As Wendell Bell has extensively argued (1997, vol. 1, p. 76), present possibilities, 'dispositionals', probabilities, beliefs about what is a desirable future, or any of a range of other ideas about the future, all exist *now*, in the present, and so therefore *can* be the subject of 'empirical' inquiry, even if the future itself cannot.[1]

Even a brief look through the futures literature reveals that the concept of 'the image of the future' is one of, if not the, key defining aspects of futures research. From the seminal work of Polak (1955; 1961; 1973; van der Helm 2005), to the related work of Boulding (1956; 1964), to the perspectives on social change of Bell and Mau (1970; 1971) and Massé (1972), to the commentary of Huber (1978), to the views of Slaughter (1991), Nandy (1996), Masini (1999), and Dator (1998; 2002), to the recent work of Rubin and collaborators (2005; 1999; 2001) – to name just a few – it seems that, if almost nothing else, one thing which most futurists would agree upon is the importance of the 'image of the future'.

Different futurists tend to focus on different aspects of images of the future – that is, not only on their nature, causes, and the consequences which stem from these images (e.g., Bell 1997, vol.1, pp. 81-6), which may include strategy formation, planning, or social and political change – but also on their deeper origins, formation, content, types, and even the role of consciousness in their characteristics. With this in mind, it would seem that a useful concise working conception of futures research might simply be that it is inquiry into (among other things, but most especially) 'images of the future' and the wide variety of inputs into, outputs from, and consequences which flow from these images of the future, in human activity and decision-making. Futurists, therefore, as those who undertake inquiry into images of the future and their many above-mentioned aspects – be it as researchers or practitioners – will of course make use of a variety of different inquiry approaches, each of which will have

1 It is of interest to note, in this regard, that empirical brain-imaging techniques have recently been used to examine the 'objective' neuro-physiological correlates of 'subjective' interior conscious experiences of forming images of the future (see Szpunar, Watson & McDermott 2007, and references therein).

fundamental philosophical assumptions and methodological commitments. This will be further explored below.

Bell has argued (1997, Ch. 1, p. 181) that, since one the main purposes of futures research is to inform decision-making and action, it can therefore "be considered an action science in the fullest sense of the term" as used in the well-known work of Argyris, Putnam and Smith (1985). Niiniluoto (2001) used the term 'design science' to capture the same idea, and other futurists have also stressed the 'action' or 'design' element of futures research (e.g., Rubin & Kaivo-Oja 1999). In short, futures research cannot be regarded as simply a pure academic conceptual exercise, disconnected from practical action in the world. Rather, it is intimately involved in the creation of the very futures which it attempts to profile through the study of images. Since images of the future influence our decision-making in the present, they are therefore complicit in the creation of the future which ultimately eventuates, and thus the study of them is also related to undertaking better-informed action. This is the 'continuous feedback loop' of future-creation, which loops 'forward' into the future as images which guide actions 'back here' in the present, and ultimately 'returns' to the future when the future becomes the present, cycling endlessly. This self-altering quality automatically distinguishes futures research (as an 'action' or 'design science') as a different class of knowledge from older and better-known sciences such as physics or chemistry. Futurist and historian Warren Wagar (1993) even argued that futures inquiry is a form of "applied history". Of course, knowledge about how the past has led to the present, how the present influences the future, how our ideas of cause and effect condition our decision-making, and how all these and other factors interplay to create the future which ultimately eventuates are also involved in human decision-making and action, and so these, too, also need to be considered alongside images of the future. A detailed "cybernetic-decisional" model of social change incorporating all of these elements was described decades ago by Bell and Mau (1970; 1971).

Because the products or outputs of futures research may have major implications for the conduct of human affairs, futures research must be undertaken as rigorously and as carefully as it possibly can be. A key part of this rigour is the conscious recognition of the extent and limits to not only the methodological approaches used, but also their philosophical bases. It is important for this reason to examine the philosophical bases upon which our statements and knowledge claims are made, including knowledge claims about the future. And it is therefore important to consider different paradigms of knowledge creation, to see what, if anything, these may illuminate about the foundations of futures research and how it has been, is, and can be, carried out.

4 A classification schema for inquiry paradigms

There are many classification schemas for inquiry paradigms, and a look at almost any book dealing with the conduct of research will reveal some sort of typology. One of the better-known classification systems is the one developed by Guba and Lincoln in various editions of the very influential *Handbook of Qualitative Research* (Denzin & Lincoln 1994; 2000; 2005). According to Guba and Lincoln (1994, p. 107):

> *A paradigm may be viewed as a set of basic beliefs (or metaphysics) that deals with ultimates or first principles. It represents a worldview that defines, for its holder, the nature of "the world", the individual's place in it, and the range of possible relationships to that world and its parts .... The beliefs are basic in the sense that they must be accepted simply on faith (however well argued); there is no way to establish their ultimate truthfulness. If there were, the philosophical debates ... would have been resolved millennia ago.*

These 'basic beliefs', which are central to the different paradigms, may be found from the answers they would give to several fundamental questions. These questions are (Guba & Lincoln 1994, p. 108):

1. the ontological question: what is the nature of 'reality' and therefore what is there that can be known?
2. the epistemological question: what is the nature of knowledge, the relationship between the would-be knower and what can be known? And,
3. the methodological question: how can the would-be knower or inquirer go about finding out whatever can be known?

To this set of three basic questions, they later added a fourth (Lincoln & Guba 2000, pp. 168-9), in response to some commentary upon and extension to their work (Heron & Reason 1997):

4. the axiological question: what is intrinsically worthwhile?

In addition, they define and examine several issues or themes which run across and through all of the classes of inquiry paradigms they consider. These themes include: the aim or purpose of the inquiry; assumptions about the nature of how knowledge accumulates; the 'voice' or 'posture' of the inquirer; the roles of values in inquiry; the criteria for assessing the quality of work; and so on. (See Table 6.2 in each of Guba & Lincoln (1994) and Lincoln & Guba (2000), and Tables 8.1-8.4 in Guba & Lincoln (2005). For convenience, some elements of these tables have been adapted and reproduced in the Appendix to this chapter, in Tables 1 and 2.)

In their view, the different answers which are given to the basic fundamental questions actually *define* an inquiry paradigm, and thence characterise the stances taken on each of the main themes or issues. They then note (Guba & Lincoln 1994, p. 112) that

differences in paradigm assumptions cannot be dismissed as mere 'philosophical' differences; implicitly or explicitly, these positions have important consequences for the practical conduct of inquiry, as well as for the interpretations of findings.

In other words, paradigmatic assumptions affect, as a result, the overall methodological approach taken, the types of methods, techniques and tools that are considered valid, and the meanings and interpretations which are assigned to the results or data that have been generated by these practices.

Guba and Lincoln considered only Western approaches to knowledge inquiry, and initially (1994) posited four major classes of inquiry paradigm which they later expanded to five (Lincoln & Guba 2000) in response to the commentary from Heron and Reason (1997) who, as mentioned above, also suggested the explicit consideration of the axiological question as foundational to paradigm definition. What is centrally important in the discussion here is not the specific details of how many inquiry paradigms there are (in the various authors' opinions), or whether they are 'Western' or 'non-Western' or of a different kind, but rather the very observation *itself* – that there *are* different inquiry paradigms, which have fundamental distinctions and differences – and that variations between them are apparent when the paradigms are examined side by side. This has many implications for understanding how the human knowledge quest has been undertaken over the course of history. It is also important for understanding how futures research has evolved over time. Other schema or typologies of forms of knowledge inquiry could equally well be used, but the overall broad shape and direction of the argument would be essentially similar, even as specific details might vary.

The five main classes of paradigm which these authors consider are:

1. positivism
2. post-positivism
3. Critical Theory and its variants, or 'criticalism'
4. constructivism, and
5. the 'participatory' paradigm

and their major features are summarised in Tables 1 and 2 in the Appendix. The commentary presented here is based on a distillation of the positions taken and observations made in the above-cited works. Drawing upon an idea of Reason and Torbert (2001), it is also sometimes useful to consider this five-part typology as consisting of three main classes: *positivistic* (positivism and post-positivism); *interpretivistic* (criticalism and constructivism); and *action/participatory*.

The first of these paradigms, positivism, represents the so-called 'received view' of scientific inquiry over the last few centuries and nowadays it most often functions primarily as the foil against which other paradigms are compared. Post-positivism arose as a result of attempts to address some of the key weaknesses which have been identified in the pure positivist viewpoint. Criticalism arose as part of the post-modernist movement of the 20th century and, to a greater or lesser degree, in opposition to the earlier positivistic paradigms. Constructivism has some features in common with criticalism, although there are significant differences between the two paradigms (see Schwandt 1994, 2000, for a detailed comparison of these positions), and the participatory paradigm introduces new assumptions, most especially about ontology and epistemology, but also in respect of almost all other foundational assumptions and issues (see Heron & Reason (1997), Reason (1994), and Reason & Bradbury (2001) for more details, and see Lincoln (2001) for a comparison between the constructivist and participatory paradigms.) What is of most interest and use to us here is to note the essential differences in the various foundational positions of the different classes of paradigm (Table 1), as well as very briefly noting in passing some of their different stances on certain issues related to knowledge inquiry (Table 2).

5 Comparison of inquiry paradigms

Looking across Table 1, we can trace a shift in the ontological positions of the five inquiry paradigms. The stances move from: a 'real', objective, external but nonetheless knowable reality in positivism; to an external objective reality which is only imperfectly knowable in post-positivism; to an historically-contingent reality in criticalism which has formed over time through the reification of initially-plastic social structures; to multiple realities in constructivism which are dependent upon the relative specifics of the particular inquiry group; to a subjective-objective participative reality literally co-created by the interaction of the inquiring consciousness and the cosmos. In the two positivistic paradigms, reality remains external to the subjectivity of the inquirer but, in the other three, reality becomes increasingly contingent upon inquirer subjectivity so that, ultimately, in the participatory paradigm, the inquirer's own subjectivity is considered to be literally *formative* of it.
We see a similar shift in the stances taken with respect to epistemology, axiology, methodology, the role of values, inquirer 'posture', and so on, and a careful reading of Tables 1 and 2 will reward the reader with many insights

into these basic issues and paradigmatic commitments. Here, for reasons of space, we shall focus most on epistemology and methodology.

The shifts in epistemological positions are especially interesting, as these of course form the basis for any knowledge claims which are produced by methodological interventions. We can see a change from the objectivist stances in the two positivistic paradigms – a view that the inquirer or would-be knower is separate and distinct from the object of knowledge ('dualism') – to the subjectivist stance taken in criticalism and constructivism – whereby knowledge is no longer considered 'objective' and therefore allegedly independent of the observer, but rather is influenced by the transaction between the would-be knower(s) and the object(s) of inquiry. In the criticalist view the findings are mediated (or 'coloured') by the value systems in operation, while constructivism takes a stronger stance and holds that the findings are *co-created* by the inquirer and the object of inquiry through the very act of inquiry itself. Both of these views assume knowledge is primarily a function of mind – knowledge claims are expressed as propositions, which latter are mental constructs (as indeed they are in the two positivistic views). In the participatory paradigm, however, this 'propositional' form of knowledge is considered only one of *four* main types of knowledge: direct 'experiential' knowledge is prior to the propositional form, as is the 'presentational' form. These three forms of knowledge are considered useful insofar as they lead to the fourth, 'practical' knowing – knowing how to *do* something, which is considered the highest form of knowledge – hence the participatory paradigm's emphasis on the primacy of 'practical knowing' (Table 2). In this view, my direct experience of the rain on my upturned face during a rain shower is also a form of knowledge, even in the absence of a theory of rainfall or climate, and is prior to any conceptual propositional knowledge I might convey to you about the experience, or any presentational form I might use to represent (i.e., 're-present') it to you, such as through metaphor, song, dance, poetry, and so on. (By way of an example: earlier in my career I was sometimes asked what it was like to do theoretical physics research. The only answer I could give which ever came close to feeling satisfactory on my part was this: "it is like the second movement of Beethoven's Ninth Symphony.")

On closer inspection, we can see in the epistemological positions of the five paradigms a three-part evolution in the emphasis placed on different forms of knowing. Following, for example, Reason and Bradbury (2001, p. xxv), Chandler and Torbert (2003), or Reason and Torbert (2001), these forms of knowledge inquiry may be termed 'first-person', 'second-person' and 'third-person', and in a similar vein, Wilber (2000, p. 70) calls them 'I' (first person), 'we/us' (second person) and 'it/its' (third person, singular and plural). As noted above, one can simplify discussions of inquiry paradigms into three main types – positivistic, interpretivistic, and action/participatory – and this maps very suggestively to what Reason and Torbert (2001) consider third-person, second-person and first-person modes of inquiry, respectively. (See also Torbert 2000, for another view of social science paradigms and first-, second- and third-person research/ practice.) In the positivistic paradigms, the emphasis is on 'objective', propositional knowledge; this is 'third-person' knowledge – the knowledge developed is about objectively-measurable qualities of material 'objects', things or 'its' (even when they are people). In the interpretivistic paradigms, the emphasis is placed on the subjective knowledge developed by a group of inquirers about some theme, issue or domain of inquiry; this is 'second-person' knowledge, as it is concerned with the shared, inter-subjective forms of knowledge which groups of people develop when they meet in a 'we' or 'us' space of discussion, dialogue, dialectic or hermeneutical meaning-making. While these two forms of knowing are also present in the participatory paradigm, it also adds the distinctly 'first-person' knowing of *direct experience*, a type of knowledge that cannot be transmitted via the mental-level constructs of propositional knowing, which latter is the basis of knowledge in the other paradigms, nor even via the 're-presentational' forms mentioned earlier. Some of the different participatory approaches, such as 'action inquiry' (Reason 1994; Torbert 2001), focus squarely on the subjectivity of the individual inquirer in the midst of action, while others, such as 'co-operative inquiry' and 'action research' (Heron & Reason 2001; Reason 1994), are more usually conducted with larger groups of people. Nonetheless the key addition to epistemological validity in this paradigm is the admission of forms of knowing which are not based solely in mental-level, conceptual propositional knowing, but which could emanate from other aspects or levels of first-person subjective human experience. And what is more, this knowing can itself be subject to critical self-reflexive inquiry ('critical subjectivity') to ensure that it is well grounded in the experiential reality upon which it is based, as well as ensuring congruence of all of the different accepted modes of knowing.

There is also a similar progression of methodologies. The positivistic paradigms undertake experimental manipulation of the exterior objective ('third-person') world in order to examine the causal dependencies of the different factors under consideration, positivism using mostly quantitative methods, post-positivism also admitting some qualitative. The emphasis moves from naïve verification of hypotheses as 'true' in the former, to attempts at falsification of hypotheses in the latter – which hypotheses must of course survive all attempts at falsification to be admitted as 'probably true' findings. In the interpretivistic paradigms, the methods are grounded in the inter-subjective (second-person) 'world' of shared subjective experience, hence the dialogical/dialectical methods of criticalism, and the hermeneutical/dialectical methods of constructivism. In the participatory paradigm, the methods involve direct participation of the (first-person) 'subjects' of the inquiry in the very process of inquiry itself, granting equal-power status (i.e. '*political* participation') to the participants, and this participation is conducted through the exchange of information via language constructs grounded in a direct, shared, first-person experiential context. Heron and Reason (2001) have therefore called this approach "research 'with', rather than 'on', people".

In the case of the axiological stance, we see how propositional knowledge as an end in itself in the two positivistic paradigms shifts to propositional knowledge becoming simply a tool for social emancipation in the two interpretivistic paradigms. In the participatory paradigm, propositional knowledge is only considered useful insofar as it contributes to *practical* knowledge about how to flourish as human beings in balance with the rest of society and the wider cosmos. Again, we can see a shift in emphasis: a move away from the distanced, 'objective expert', 'disinterested scientist' stance or posture of the two positivistic paradigms, to a progressively more intimate engagement with the world, as an activist and advocate (criticalism), as a passionate participant/facilitator (constructivism), to a self-reflexive actor-agent engaging with others in multiple forms of knowing, knowledge-creation, and reality-creation (participatory).

6 Futures research methods through the paradigms

As will be clear to anyone who examines it, examples of all of these paradigmatic approaches can be found in the futures literature. Space does not here permit a detailed exposition of how futures methods have been influenced by the above-mentioned paradigms of inquiry. Instead, we can only give a few representative examples, and leave it to the reader to compare the paradigmatic positions described above with the wider futures research literature.

As several commentators have observed, futures research methods have undergone an evolution over the preceding several decades. Slaughter (2002), for example, writes of there having been three or four major phases in the evolution of futures methods. In his view, these have essentially been forecasting, scenarios, social construction, and, most recently, 'integral' methods (Slaughter 2004). Inayatullah (2002b) has suggested an analogous change in futures methods which, in his view, have moved from forecasting to anticipatory action learning. The former is expert-based and largely positivistic-empiricist; the latter has, as the name suggests, strong resonances with the action/participatory paradigm. A similar progression from expert-led quantitative methods to more qualitative and participatory methods can also be seen in the futures methods described by Bell (1997, vol.1, ch.6), and the mix of approaches is discernible in the variety of methods and techniques discussed in the resource collections edited by Glenn and Gordon (2003), and Slaughter, Inayatullah and Ramos (2005)

For example, the (post)positivistic paradigmatic commitment of Olaf Helmer (one of the inventors of Delphi) is apparent in his statement that "most of futures research may be regarded as a subfield of operations research" (Helmer 1983, p. 83), which latter is an archetypal rational-quantitative discipline, although he did allow for definitions of futures research which were broader than this (p. 83).

It was recognised by the mid 1970s that strongly positivistic approaches to futures research were on shaky methodological grounds, as pointed out by Ida Hoos (1978), as well as Roy Amara (1978), who noted: "the familiar tools of scientific investigation can be applied only in their most primitive forms" (p. 41). The limitations of positivistic approaches to futures research forms the essential core theme of the book edited by Linstone and Simmonds (1977), wherein the role of worldviews in futures research is seen to come to the fore. These editors succinctly characterised the crisis they perceived in futures research at that time as (p.xv):

> *No longer are we just dealing with methodological issues but with challenges to long-accepted paradigms. ... [There is a] growing awareness of the influence of the personality, experience, and character of those doing futures work, those requesting futures work, and the organizational and institutional environment in the selection of issues chosen to study. ... The heart of the matter is the perceptual change in the research worker himself.*

The 'prospective' approach of Gaston Berger (Cournand & Lévy 1973) emphasised the role of discussion and dialogue to determine what futures could be created and which of these were worth creating, which clearly demonstrates some of the paradigmatic commitments of both criticalism and constructivism, including the explicit consideration of values as intrinsic to inquiry. The idea that reality is 'socially constructed' (Berger & Luckmann 1966) also lies at the heart of Bertrand de Jouvenel's perspective on futures inquiry and informed political action (de Jouvenel 1967; Gamba 2003). One can find this interpretivist commitment to inquiry in a good deal of the Western European tradition of futures research (see, e.g., Masini 1993, 1999).

The issue of dissent and the use of dialectic is a central element of criticalist and constructivist approaches, and a special issue of the journal *Futures* focussed explicitly on the role of dissent in futures studies (Sardar 1999a) while the book edited by Sardar (1999b) is similarly dissenting and dialectical in tone and timbre. More recently, the use of action/participative methods in futures inquiry, in particular 'action research', has also been the subject of a special issue of *Futures* (Ramos 2006).

If we recall that both positivist and post-positivist approaches share the same basic aim (cf. Table 2), then we can consider there to be four main purposes of the five main classes of inquiry paradigm described above: prediction and control; critique and transformation (leading to emancipation); understanding and insight (leading to re-construction of prior constructions); and human flourishing (through political participation). Given this, we can see strong resonances of these four inquiry aims in the four types of futures approaches discussed by Inayatullah (2002a, ch.1): predictive; critical; interpretive; and anticipatory action learning.

Finally, it is worth mentioning very briefly that another approach to futures research has begun forming in recent times, based upon the principle of seeking to integrate the many different – and indeed, competing and sometimes even antagonistic – approaches to futures research which have gone before into an overall 'integrating' or 'integral' approach (see, e.g., Hines 2004; Slaughter 2004). This type of approach is not bound to any single paradigm-based perspective, but rather seeks to use the best and most appropriate aspects of all existing paradigms, depending on the nature and domain of any particular inquiry being undertaken. At the time of writing, a special issue of *Futures* dedicated to 'Integral Futures' is in press, and shows a snapshot of the current state of development and thinking in this newly-emerging approach to futures research.

The purpose of this very brief review was to highlight that all of the inquiry paradigms described above have been used to undertake futures research over the past few decades. In all such work, however, the choice of inquiry paradigm must be appropriate to the domain of inquiry. As Linstone and Simmonds (1977) found, to use just one example, the empiricist-positivistic approaches of conventional science – perfect for third-person, objective, 'it' knowledge – cannot be used to properly study the second-person, inter-subjective, 'we' realm of meaning-making and worldviews in a future-creating social system of conscious agents. To do so is a category error. And it is precisely this ability to notice such category errors, in knowledge inquiry in general and in futures research in particular, which was the main point of undertaking our careful study of the philosophical and paradigmatic foundations of knowledge inquiry and futures research.

7 Conclusion

In this chapter a typology of inquiry paradigms was examined and discussed in order to understand how these paradigms differ in their foundational assumptions, as well as how they have been used as a basis for futures research over the last several decades. It was argued that the philosophical bases of inquiry – the foundational assumptions and fundamental presuppositions about reality, knowledge and method, to name but a few – must be explicitly taken into account in order to ensure that the form and approach of an inquiry is appropriate to the purpose and domain of the inquiry. Futures research – which, as an 'action science' form of knowledge inquiry, takes as one of its primary domains the subjective realm of 'images of the future', and which seeks to not only make knowledge claims about the future but also to shape it – is especially beholden to demonstrate careful, rigorous and disciplined thinking. If futures researchers can successfully demonstrate this care and rigour with respect to the philosophical and methodological foundations of futures research and the knowledge claims which it attempts to make, then we may find an increasing receptivity to the idea of thinking seriously about the future.

Appendix

The various paradigms' basic positions on the foundational issues of ontology, epistemology, methodology and axiology are shown in Table 1, while their stances on a variety of other issues are shown in Table 2. These Tables are based on a distillation of the positions taken and observations made in Guba and Lincoln (1994; 2005), Heron and Reason (1997) and Lincoln and Guba (2000).

Table 1 Foundational stances of the five inquiry paradigms.
Adapted and distilled from Guba and Lincoln (1994; 2005), Heron and Reason (1997) and Lincoln and Guba (2000).

	Positivism	Post-positivism	Criticalism	Constructivism	Participatory
Ontology	naïve realism – 'real' reality but apprehendable	critical realism – 'real' reality but only imperfectly and probabilistically apprehenable	historical realism – virtual reality shaped by social, political, cultural, economic, ethnic and gender values; crystallised over time	relativism – local and specific co-constructed realities	participatory reality – subjective-objective reality, co-created by mind and given cosmos
Epistemology	dualist / objectivist; findings 'true'	modified dualist / objectivist; critical tradition / community; findings 'probably true'	transactional / subjectivist; value-mediated findings	transactional / subjectivist; co-created findings	critical subjectivity in participatory transaction with cosmos; extended epistemology of experiential, presentational, propositional, and practical knowing; co-created findings
Methodology	experimental / manipulative; verification of hypotheses; chiefly quantitative methods	modified experimental / manipulative; critical multiplism; falsification of hypotheses; may include qualitative methods	dialogic / dialectical	hermeneutical / dialectical	political participation in collaborative action inquiry; primacy of the practical; use of language grounded in shared experiential context
Axiology	propositional knowing about the world is an end in itself, is intrinsically valuable		propositional, transactional knowing is instrumentally valuable as a means to social emancipation, which is an end in itself, is intrinsically valuable		practical knowing how to flourish with a balance of autonomy, cooperation, and hierarchy in a culture is an end in itself, is intrinsically valuable

Table 2 Paradigm positions on selected issues.
Adapted and distilled from Guba and Lincoln (1994; 2005), Heron and Reason (1997) and Lincoln and Guba (2000).

	Positivism	Post-positivism	Criticalism	Constructivism	Participatory
Inquiry aim	explanation: prediction and control		critique and transformation; restitution and emancipation	understanding; reconstruction	human flourishing
Inquirer posture	'disinterested scientist' as informer of decision makers and change agents		'transformative intellectual' as advocate and activist	'passionate participant' as facilitator of multi-voice reconstruction	primary voice manifest through aware self-reflective action; secondary voices in illuminating theory, narrative, movement, song, dance, and other presentational forms
Nature of knowledge	verified hypotheses established as facts or laws	non-falsified hypotheses that are probable facts or laws	structural / historical insights	individual or collective reconstructions sometimes coalescing around consensus	extended epistemology; primacy of practical knowing; critical subjectivity; living knowledge
Knowledge accumulation	accretion – 'building blocks' adding to 'edifice of knowledge'; generalisations and cause-and-effect linkages		historical revisionism; generalisation by similarity	more informed and sophisticated reconstructions; vicarious experience	in communities of inquiry embedded in communities of practice
Values	excluded – influence denied; considered to be extrinsic to inquiry		included – formative; considered to be intrinsic to inquiry		
Goodness or quality criteria	conventional benchmarks of 'rigour'; internal and external validity, reliability and objectivity		historical situatedness; erosion of ignorance and misapprehensions; action stimulus	trustworthiness and authenticity including catalyst for action	congruence of experiential, presentational, propositional and practical knowing; leads to action to transform the world in the service of human flourishing

References

Amara, R. 1974, 'The futures field: functions, forms, and critical issues', *Futures*, vol. 6, no. 4, pp. 289-301.

1978, 'Probing the future', in *Handbook of futures research*, ed. J. Fowles, Greenwood Press, Westport, CT, USA, pp. 41-51.

1981, 'The futures field: searching for definitions and boundaries', *The Futurist*, vol. 15, no. 1, pp. 25-9.

Argyris, C., Putnam, R. & Smith, D.M. 1985, *Action science: concepts, methods, and skills for research and intervention*, Jossey-Bass, San Francisco.

Bell, W. 1997, *Foundations of futures studies*, 2 vols, Transaction Publishers, New Brunswick, NJ, USA. vol.1 – History, Purposes, Knowledge; vol.2 – Values, Objectivity, and the Good Society.

2002a, 'Advancing futures studies: a reply to Michael Marien', *Futures*, vol. 34, no. 5, pp. 435-47.

2002b, 'A community of futurists and the state of the futures field', *Futures*, vol. 34, no. 3-4, pp. 235-47.

2005, 'An overview of futures studies', in *The knowledge base of futures studies*, Professional edn, eds R. A. Slaughter, S. Inayatullah & J. M. Ramos, Foresight International, Brisbane, Australia.

Bell, W. & Mau, J.A. 1970, 'Images of the future: theory and research strategies', in *Theoretical sociology: perspectives and developments*, eds J. C. McKinney & E.A. Tiryakian, Appleton-Century-Crofts, New York, pp. 205-34.

1971, 'Images of the future: theory and research strategies', in *The sociology of the future: theory, cases, and annotated bibliography*, eds W. Bell & J.A. Mau, Russell Sage Foundation, New York, pp. 6-44.

Berger, P.L. & Luckmann, T. 1966, *The social construction of reality: a treatise in the sociology of knowledge*, 1991 reprint edn, Penguin.

Boulding, K.E. 1956, *The image: knowledge in life and society*, The University of Michigan Press, Ann Arbor, MI, USA.

1964, *The meaning of the twentieth century: the great transition*, Harper.

Chandler, D. & Torbert, W.R. 2003, 'Transforming inquiry and action: Interweaving 27 flavors of action research', *Action Research*, vol. 1, no. 2, pp. 133-52.

Cournand, A. & Lévy, M. (eds) 1973, *Shaping the future: Gaston Berger and the concept of Prospective*, Current topics of contemporary thought, vol. 11, Gordon and Breach, New York.

Dator, J.A. 1998, 'The future lies behind! Thirty years of teaching future studies', *American Behavioral Scientist*, vol. 42, no. 3, pp. 298-319.

2002, 'The future lies behind! – thirty years of teaching futures studies', in *Advancing futures: futures studies in higher education*, ed. J.A. Dator, Praeger Pubs., Westport, CT, USA, pp. 1-30.

2005, 'Foreword', in *The knowledge base of futures studies*, Professional edn, eds R. A. Slaughter, S. Inayatullah & J.M. Ramos, Foresight International, Brisbane, Australia.

de Jouvenel, B. 1967, *The art of conjecture*, trans. N. Lary, Weidenfeld and Nicholson, London.

Denzin, N.K. & Lincoln, Y.S. (eds) 1994, *Handbook of qualitative research*, Sage Pubs., Thousand Oaks, CA, USA.

(eds) 2000, *Handbook of qualitative research*, 2nd edn, Sage Pubs., Thousand Oaks, CA, USA.

(eds) 2005, *The SAGE handbook of qualitative research*, 3rd edn, Sage Pubs., Thousand Oaks, CA, USA.

Einstein, A. to R.A. Thornton, unpublished letter dated 7 December 1944, EA 61-574, Einstein Archive, Hebrew University, Jerusalem.

Gamba, G. 2003, *Prediction, power and good governance: according to Bertrand de Jouvenel*, trans. M. Brady, Institute of International Sociology, Gorizia, Italy. Dual-language booklet in two inverted halves. Italian and English.

Glenn, J.C. & Gordon, T.J. (eds) 2003, *Futures research methodology*, CD-ROM version 2.0 edn, American Council for the United Nations University, Washington, DC.

Godet, M. 2000, 'How to be rigorous with scenario planning', *Foresight*, vol. 2, no. 1, pp. 5-9.

Guba, E.G. & Lincoln, Y.S. 1994, 'Competing paradigms in qualitative research', in *Handbook of qualitative research*, eds N.K. Denzin & Y.S. Lincoln, Sage Pubs., Thousand Oaks, CA, USA, pp. 105-17.

2005, 'Paradigmatic controversies, contradictions, and emerging confluences', in *The SAGE handbook of qualitative research*, 3rd edn, eds N. K. Denzin & Y.S. Lincoln, Sage Pubs., Thousand Oaks, CA, USA, pp. 191-215.

Harman, W.W. 1976, *An incomplete guide to the future*, San Francisco Book Company, Inc.

Helmer, O. 1983, *Looking forward: a guide to futures research*, Sage Publications, Beverly Hills, CA, USA.

Heron, J. & Reason, P. 1997, 'A participatory inquiry paradigm', *Qualitative Inquiry*, vol. 3, no. 3, pp. 274-94.

2001, 'The practice of co-operative inquiry: research 'with' rather than 'on' people', in *Handbook of action research: participative inquiry and practice*, eds P. Reason & H. Bradbury, Sage Pubs., Thousand Oaks, CA, USA, pp. 179-88.

Hines, A. 2004, 'Applying integral futures to environmental scanning', *Futures Research Quarterly*, vol. 19, no. 4, pp. 49-62.

Hoos, I.R. 1978, 'Methodological shortcomings in futures research', in *Handbook of futures research*, ed. J. Fowles, Greenwood Press, Westport, CT, USA, pp. 53-66.

Howard, D.A. 2004, 'Einstein's philosophy of science', *The Stanford Encyclopedia of Philosophy (Spring 2004 edn)*, ed. E.N. Zalta, Stanford University, Stanford, CA, USA, <http://plato.stanford.edu/archives/spr2004/entries/einstein-philscience/>.

Huber, B.J. 1978, 'Images of the future', in *Handbook of futures research*, ed. J. Fowles, Greenwood Press, Westport, CT, USA, pp. 179-224.

Inayatullah, S. 2002a, *Questioning the future: futures studies, action learning and organisational transformation*, Tamkang University, Taipei, Taiwan.

2002b, 'Reductionism or layered complexity? The futures of futures studies', *Futures*, vol. 34, no. 3-4, pp. 295-302.

Lincoln, Y.S. 2001, 'Engaging sympathies: relationships between action research and social constructivism', in *Handbook of action research: participative inquiry and practice*, eds P. Reason & H. Bradbury, Sage Pubs., Thousand Oaks, CA, USA, pp. 124-32.

Lincoln, Y.S. & Guba, E.G. 2000, 'Paradigmatic controversies, contradictions, and emerging confluences', in *Handbook of qualitative research*, 2nd edn, eds N.K. Denzin & Y.S. Lincoln, Sage Pubs., Thousand Oaks, CA, USA, pp. 163-88.

Linstone, H.A. & Simmonds, W.H.C. (eds) 1977, *Futures research: new directions*, Addison-Wesley, Reading, MA, USA.

Marien, M. 2002a, 'Futures studies in the 21st Century: a reality-based view', *Futures*, vol. 34, no. 3-4, pp. 261-81.

2002b, 'My differences with Wendell Bell', *Futures*, vol. 34, no. 5, pp. 449-56.

Masini, E.B. 1993, *Why futures studies?*, Grey Seal, London.

1999, 'Rethinking futures studies', in *Rescuing all our futures: the future of futures studies*, ed. Z. Sardar, Praeger, Westport, CT, pp. 36-48.

Massé, P. 1972, 'Attitudes towards the future and their influence on the present', *Futures*, vol. 4, no. 1, pp. 24-9.

Michael, D.N. 1985, 'With both feet planted firmly in mid-air: reflections on thinking about the future', *Futures*, vol. 17, no. 2, pp. 94-103.

Nandy, A. 1996, 'Bearing witness to the future', *Futures*, vol. 28, no. 6-7, pp. 636-9.

Niiniluoto, I. 2001, 'Futures studies: science or art?' *Futures*, vol. 33, no. 5, pp. 371-7.

Polak, F.L. 1955, *De toekomst is verleden tijd* [The future is past tense], 2 vols, W. de Haan, Utrecht, The Netherlands.

1961, *The image of the future: enlightening the past, orienting the present, forecasting the future*, 2 vols, trans. E. Boulding, Oceana Pubs, New York.

1973, *The image of the future*, trans. E. Boulding, Elsevier Scientific Publishing, Amsterdam. Translation of the abridged work: Dutch title: De toekomst is verleden tijd (The future is past tense). Hilversum: W. de Haan, 1968. Original two-volume work published in 1955. An electronic version has been put on-line at <http://www.cnam.fr/lipsor/eng/memoryofprospective.php>.

Ramos, J.M. 2006, 'Action research and futures studies', *Futures*, vol. 38, no. 6, pp. 639-41.

Reason, P. 1994, 'Three approaches to participative inquiry', in *Handbook of qualitative research*, eds N.K. Denzin & Y.S. Lincoln, Sage Pubs., Thousand Oaks, CA, USA, pp. 324-39.

Reason, P. & Bradbury, H. (eds) 2001, *Handbook of action research: participative inquiry and practice*, Sage Pubs., Thousand Oaks, USA.

Reason, P. & Torbert, W.R. 2001, 'The action turn: towards a transformational social science', *Concepts and Transformation*, vol. 6, no. 3, pp. 1-37.

Rubin, A. 2005, 'How to study images of the future', in *The knowledge base of futures studies*, Professional edn, eds R.A. Slaughter, S. Inayatullah & J.M. Ramos, Foresight International, Brisbane, Australia, vol. 2.

Rubin, A. & Kaivo-Oja, J.Y. 1999, 'Towards a futures-oriented sociology', *International Review of Sociology*, vol. 9, no. 3, pp. 349-71.

Rubin, A. & Linturi, H. 2001, 'Transition in the making. The images of the future in education and decision-making', *Futures*, vol. 33, no. 3-4, pp. 267-305.

Sardar, Z. 1999a, 'Dissenting futures and dissent in the future', *Futures*, vol. 31, no. 2, pp. 139-46.

(ed.) 1999b, *Rescuing all our futures: the future of futures studies*, Praeger Studies on the 21st Century, Praeger, Westport, CT.

Schwandt, T.A. 1994, 'Constructivist, interpretivist approaches to human inquiry', in *Handbook of qualitative research*, eds N. K. Denzin & Y. S. Lincoln, Sage Pubs., Thousand Oaks, CA, USA, pp. 118-37.

2000, 'Three epistemological stances for qualitative enquiry: interpretism, hermeneutics, and social constructionism', in *Handbook of qualitative research*, 2nd edn, eds N.K. Denzin & Y.S. Lincoln, Sage Pubs., Thousand Oaks, CA, USA, pp. 189-213.

Slaughter, R.A. 1991, 'Changing images of futures in the 20th century', *Futures*, vol. 23, no. 5, pp. 499-515.

2002, 'From forecasting and scenarios to social construction: changing methodological paradigms in futures studies', *Foresight*, vol. 4, no. 3, pp. 26-31.

2004, 'Towards integral futures', in *Futures beyond dystopia: creating social foresight*, RoutledgeFalmer, London, pp. 152-66.

Slaughter, R.A., Inayatullah, S. & Ramos, J. M. (eds) 2005, *The knowledge base of futures studies*, 5 vols, CD-ROM Professional edn, Foresight International, Brisbane, Australia.

Szpunar, K.K., Watson, J.M. & McDermott, K.B. 2007, 'Neural substrates of envisioning the future', *Proceedings of the National Academy of Sciences USA*, vol. 104, no. 2, pp. 642-7.

Torbert, W.R. 2000, 'Transforming social science: integrating quantitative, qualitative, and action research', in *Transforming social inquiry, transforming social action: New paradigms for crossing the theory/practice divide in universities and communities*, eds F.T. Sherman & W.R. Torbert, Kluwer / Springer, Amsterdam / Berlin, pp. 67-91.

2001, 'The practice of action inquiry', in *Handbook of action research: participative inquiry and practice*, eds P. Reason & H. Bradbury, Sage Pubs., Thousand Oaks, CA, USA, pp. 250-60.

van der Helm, R. 2005, 'The future according to Frederik Lodewijk Polak: finding the roots of contemporary futures studies', *Futures*, vol. 37, no. 6, pp. 505-19.

Wagar, W.W. 1993, 'Embracing change: futures inquiry as applied history', *Futures*, vol. 25, no. 4, pp. 449-55.

Wilber, K. 2000, *A theory of everything: an integral vision for business, politics, science and spirituality*, Shambhala Pubs., Boston.

6 Macrohistory, macrohistorians and futures studies

Sohail Inayatullah

1 Introduction

Through its delineation of the patterns of history, macrohistory gives a structure to the often fanciful visions of futurists. Macrohistory gives us the weight of history balancing the pull of the image of the future. Yet like futures studies, it seeks to transform past, present and future not merely reflect upon social space and time.

Based on a special issues of *New Renaissance* (1996, vol. 7, no. 1) and the book *Macrohistory and Macrohistorians* (Galtung & Inayatullah, 1997; see also Hayward and Voros, 2006; Voros, 2006; Inayatullah, 2004a), this essay links macrohistory with futures studies. It takes the views of twenty or so macrohistorians and asks what they offer to the study of alternative futures. The macrohistorians used for this article include: Ssu-Ma Ch'ien, St. Augustine, Ibn Khaldun, Giambatista Vico, Adam Smith, G.W.F. Hegel, Auguste Comte, Karl Marx, Herbert Spencer, Vilfredo Pareto, Gaetano Mosca, Max Weber, Oswald Spengler, Teilhard de Chardin, Pitirim Sorokin, Arnold Toynbee, Rudolf Steiner, Fernand Braudel, Fred Polak, Nikolai Kardashev, Prabhat Rainjan Sarkar, Elise Boulding, Riane Eisler, Johan Galtung and Gaia herself (as articulated by James Lovelock).

2 History of social systems

Macrohistory is the study of the histories of social systems, along separate trajectories, through space and time, in search of patterns, even laws of social change. Macrohistory is thus nomothetic (generalized understanding, searching for similarities) and diachronic (through long periods of time). Macrohistorians – those who write macrohistory – are to the historian what an Einstein is to the run-of-the-mill physicist: in search of the totality of space and time, social or physical. Macrohistorians use the detailed data (local, national and even world histories) of historians for their grand theories of individual, social, civilizational and global change.

Writes Johan Galtung:

> *Much is required of the macro-historian. He or she should have a command of the basic empirical features of some single cases, contiguous/continuous or not. The mind has to survey vast terrains in space-time. A high capacity for pattern recognition is a precondition. The macrohistorian has to make sense of enormous amounts of data. Whether the focus is on the long term trajectory of one unit, for instance a country, or the more ambitious comparison of several trajectories, patterns have to emerge, be imposed on the data, the degree of correspondence between theory and data has to be evaluated, the patterns have to be revised, etc. The macrohistorian is not producing a diachrony of one unit (historians do) or (more interesting) parallel diachronies of several units, whether organized in tables/charts or in historical atlases. He or she is looking for recurring patterns, in the trajectories of the same and/or different units, and for mechanisms underlying them (Galtung and Inayatullah, 1997, p. 4).*

We can see that in some ways this is required of the futurist as well. She or he must have high pattern recognition and must have a theory of change. A theory of change is different from a typology, as it explains why there is movement from one type to another – the causes and mechanisms of change – and what is the underlying nature/shape of change.

Macrohistorians and their macrohistories have much to offer futures studies. Futures studies, while strong at challenging the present, is often weak at contouring the parameters of the future possible. Macrohistory through its delineation of the structures of history – of the causes and mechanisms of historical change; of inquiry into what changes and what stays stable; of an analysis of the units of history; and a presentation of the stages of history – provides a structure from which to forecast and gain insight into the future.

By knowing what historically can and cannot change, scenarios of the future can be more plausible. By exploring the range of units or collectivities, we can break out of the straitjacket of nations as our only unit for the future. Finally by understanding the stages of history, we can better understand the stages of the future.

Macrohistory gives us the weight of history balancing the pull of the image of the future. It gives a historical distance to the many claims of paradigm shifts, allowing us to distinguish between what are mere perturbations and what are

genuine historical transformations. While giving us insights into the human condition, macrohistory also intends to explain past, present and future, and to a certain extent forecast the movement of units through time.
Indeed, I have argued elsewhere (Inayatullah 2002a) that macrohistory should be considered one of the six pillars of Futures Studies (MATDCT). These pillars include: (1) Mapping the future – understanding trends, images and weights of the unit under study; (2) Anticipating the future – understanding emerging issues and trends; (3) Timing the future – understanding the multiple shapes of time (linear, cyclical, spiral, bifurcation, for example), that is, macrohistory; (4) Deepening the future – understanding the future from the viewpoint of different worldviews and myths; (5) Creating alternatives – using scenarios to reduce uncertainty and imagine alternatives; and (6) Transforming the future – using visioning (imagining the preferred future), backcasting and action learning to invent a new future.

3 Historical and epistemic context

As with futurists, as macrohistorians make claims of the empirical, it is important to locate them within the historical conditions they write in and the episteme that frames what is knowable. For example, Islamic philosopher Ibn Khaldun asserts that his sociology is time, space and culture invariant, yet the bases for his theory emerged from the primacy of unity in Islamic thought. His theory of history and his comments on historiography can be located within the current thinking of the time, the 14th century. Given his knowledge of Bedouins, personal participation in various coups and assorted palace intrigues, and the centrality of unity in Islamic cosmology, it is not surprising that what emerges is a history of dynastic change with Bedouin challenges to "civilization" leading to unity employed as the key explanatory variable.
This is true for Marx as well. Although Marx attempted to create a perfect new world realizable through an objective understanding of the real, he was responding to a tradition as well – the concerns of nineteenth century Europe. His thinking was contextualised by the rationality of the Enlightenment and its German response (the idealistic perspective of Kant and Hegel). Given the idealistic nature of the philosophical nexus around him and of the recent Christian past, he claimed that his work was a science of the objective of the material world, and not a speculation on the idealistic or religious world of the medieval era. And given the rapidity of technological change today, it is

not surprising how many futurists focus on technology as the prime mover of change, seeing it as leading to new social stages (Toffler, for example). But, we are often unaware of the paradigm, the context, we write in.

Much earlier, perhaps the first macrohistorian, Ssu-Ma Ch'ien (145-90BC), also faced the problem of context. Chinese macrohistorian Ssu-Ma Ch'ien had to write within his episteme before he could transform it. To be intelligible, he had to write in the context of Confucian history. There were various problems to be addressed such as the periodicity of history, the causes of the rise and fall of dynasties, the role of the Tao, the importance of morality in giving meaning to history, and the centrality of the golden past. However unique and revolutionary his writing was, it still existed within a particular sea of knowledge.

Futures studies as well exists within a particular frame – that of the problem of modernity and postmodernity (dramatic new technologies and diversity of new worldviews). In response to the needs of capitalism and the inter-state system, futures studies has sought to develop predictive models of economy and national security. At the same time, a counter hegemonic discourse has developed tied to deconstructing dominant forms of knowledge and creating dissenting futures.

Along with the historical context of the macrohistorian, crucial to understanding the future are the stages of history posited. Comte had his theological (based on religion-faith), metaphysical (based on philosophy-reason) and positive (based on science-truth). Sorokin has his three ages of the ideational, dualistic-integrated and sensate but with a fourth stage as the transition, the age of scepticism and chaos.[1] Spencer relates his societal types to phases in history: barbarism, militant, industrial, and a fourth yet to emerge. Vico has his Age of Gods, Heroes, Men and Barbarians (from which we return to the Age of Gods) and Ibn Khaldun argues for a primitive-civilization-primitive pattern.3. Agency, structure and the transcendental

Along with stages are issues as to the role of choice in creating history and future. Most social theorists argue back and forth between agency and structure. However, macrohistorians find escape ways out of these categories. For example, for Vico, history and future, although patterned, are not predetermined – there are laws but these are soft. As Attila Faj writes:

1 Sorokin devises his stages from empirical data and from asking the question what is real. The answers are (1) matter is real, (2) mind is real, (3) both are real, (4) nothing is real, (5) one cannot know. Responses 4 and 5, even if true, cannot lead to a meaningful society. The other three can, creating the sensate, ideational and integrated society. Change occurs since human beings have a richer spectrum of needs than any particular social formation can satisfy. Once society reaches the sensate gutter ie from ideational to integrated to sensate, then the desire for guidance, for the fulfilment of other needs pulls it upward again. Until the next swing.

> *The famous corsi and ricorsi are both rheological and chorological, that is, circling "softly," round dancing. The softness of the law means that the successive figures of this roundelay are not necessarily unavoidable and are not independent of any condition and circumstance. Each historical stage streams into the following one and gets mixed with it, so we cannot distinguish them sharply. For a long stretch, the stages and everything that belongs to them are mingled like the sweet water of an estuary with the salt water of the sea (Faj, 1987, pp. 22-23).*

Critics, however, point out that macrohistory by focusing on the grand stages – the laws of history – removes choice and contingency, privileges structure over human agency and misses too many significant details. However, while the structure/agency dilemma is central within the linear/developmentalist model or the cyclical/fatalistic model, but, among others, Sarkar (*varna* – collective psychology/types of power), Galtung (cosmology), Foucault (discourse) and Sorokin (supersystems – sensate, idealist/integrative and ideational) give us ways out of these dilemmas. For example, for Sarkar there is historical structure (evolutionary derived), but there is individual will *and* there is a cosmic will: a grander intelligence. These exist in dialectical tension.

Privileging one perspective (agency) results in individualism or liberalism (Smith, for example). Privileging another (structure) results in structuralism (Marx, for example). And, if one moves toward the third, then divinity results (Augustine or Steiner, for example). The real has different levels – the task is to exist in them simultaneously, to develop a theory that has linear, cyclical and transcendental dimensions and has agency, structure and superagency (the transcendental) as to what causes movement through history. A theory of the future would equally need to embrace these multiple perspectives rather than mistakenly focus on any particular approach.

Choosing structure over agency would be a mistake as would be choosing agency over structure. Keeping divinity and other mysterious factors out of the macro analysis would also be a mistake. There is no necessity to make a decision to privilege a particular way of understanding; all levels of interpretation must be held on to simultaneously.

For Galtung and Foucault as well these are false choices. A particular cosmology and discourse gives us the possibility of including both horns of the structure/agency dilemma (and divinity if need be). Similarly, Sorokin develops his theory arguing that any system must have its own inner dynamics and must interact with the structure of the external world that causes external and internal change, thus allowing both agency and structure. To Galtung, Foucault and Sorokin, however, superagency is not a possibility. The intervention of God or other mysterious spiritual forces is not an empirical possibility, but rather the type of approach one gets during ideational eras, or pertaining to a particular cosmology or discourse.

For most macrohistorians, individuals are important but they exist in larger fields that condition their choices: epistemological, ontological, economic and cultural or class, gender, *varna*, civilization type, dynasty, cultural personality or ways of knowing the real. Futurists, in general, tend to focus on the individual's ability in creating the future and the values that inform the good society, vision, in question. But for the macrohistorian, these value preferences in themselves exist within certain structures: biological (the evolution of the species and the environment), epistemological (the historical possibilities of what is knowable and thinkable), social (one's own culture and its history), technological (the material and social ways through which actions can be expressed), and the economic (basic needs and growth, the realities of the material world).

While some macrohistorians find ways to balance the individual and the social, others focus on the system as a whole. In the macro view, size, structure (for example, vertical/horizontal or feudal/bureaucratic arrangements), relations (person to person; person to nature; person to society) are significant and primary over individual choices and hopes. Transcendental theories, in general, focus on the individual and his (and sometimes her) relationship to the Transcendental and less on the social structures in question as, for example, in the case of Teilhard De Chardin's work.

4 Cyclical and linear

As important as tension between agency and structure is the debate between cyclical and linear schools of history. Cyclical theory privileges perpetual change while linear theory privileges equilibrium, although it could be an evolutionary equilibrium as in the case of Spencer. In cyclical theories change is endemic to the system: through dialectics, the principle of limits (wherein a historical stage by exaggerating its own nature and denying others is surpassed by another), through the Chinese yin/yang principle, or through the Indian Tantric *vidya/avidya* (introversion and extroversion) principle.

Ssu-Ma Ch'ien links the yin/yang principle with the rise and fall of the dynasty. Sarkar links *vidya/avidya* with his four stages of history[2] and with an individual's own spiritual struggle to gain *moksa* (enlightenment).

2 Sarkar believes history is the cyclical rotation of four collective psychologies – the worker, warrior, the intellectual and the merchant. Each era dialectically emerges from the previous one. In the final merchant phase, there is massive exploitation of the other psychologies, leading to a worker's revolution.

In contrast, in linear theories change is often due to external causes. Cyclical historians examine the rise and fall of civilizations. Linear historians, on the other hand, believe the fall problem applies to other civilizations (Oriental civilization, for example) while their own civilization (the West) is destined for eternal rise and progress. The formula for progress has been found; the problem now is merely staying the course. Khaldun, however, writing during the decline of Islamic civilization could see that the cause of history had to be strengthening and weakening of unity, not just the former or the latter.
While cyclical theorists have linear dimensions (the move up or they move down), it is the return to a previous stage – however modified – that does not allow for an unbridled theory of progress.
Linear theorists also have cyclical dimension to their theories. Within the narrative of linear stages, linear theorists often postulate ups and downs of lesser unit of analysis (for example, within human evolution or the evolution of capital, there might be a rise and fall of nations, firms or dynasties), but in general the larger pattern is progress. Humans might have contradictions (based on the Augustinian good/evil pattern) but society marches on either through technology, capital accumulation, innovation, the intervention or pull of God.
Spiral theorists attempt to include both, having certain dimensions which move forward and certain dimensions that repeat. Spiral theories are fundamentally about a dynamic balance.
In contrast to these approaches, Riane Eisler argues for a multi-linear approach to macrohistory. She writes:

> *Cultural transformation... is a multilinear rather than unilinear theory of early human cultural evolution: one that proposes that it did not follow a single path, but rather a variety of paths – with groups in different environments evolving in different directions, some orienting primarily to a partnership or gylanic model and others to an androcratic model (Eisler 1995, p. 1; see also Eisler 1997).*[3]

Grand thinkers (macrohistorians and futurists) are rarely useful to the present – the status quo – as they represent change. By identifying endogenous causes of change, they remind that the present is bound to transform, it cannot stay stable, nor can the idealized future society – the utopia or eutopia – for that too will transform once realized.

3 According to Eisler, "Cultural transformation theory proposes that history is neither linear, cyclical, nor purely random, but the outcome of the interaction of two types of movements. The first is the tendency of social systems to move from less to more complex forms of organization largely due to technological breakthroughs or phase changes. The second is the movement of cultural shifts between two basic models or "attractors" for social and ideological organization which I have called the dominator and partnership models – or more specifically, androcracy and gylany" (1995, p. 1).

Cyclical theories in particular are seen as pessimistic by the elite of the core nation and the core civilization. From the view of the individual, cyclical theorists are seen as disempowering since structure and process prevail over agency. Transcendental theories are empowering in that they inspire individuals to act but they also lead to fatalism since all is in the hands of the transcendental. When combined with nationalism, transcendental theories can lead to a fascism of sorts, with the rise of one's own nation guaranteed because of historical destiny – because the *geist* (spirit) has entered the nation or a world leader (as with Hegel and Sri Aurobindo).

Most macrohistorians have linear and cyclical dimensions in their theories. Following Khaldun, Weber, too, in some of his writings has a cyclical dimension. Once a historical structure is exhausted, "a charismatic leader emerges outside the structure and gives it its *coup de grace*" (Etzioni & Etzioni-Halevy 1973, p. 4). As charisma becomes routinized, the new structure will also face declining legitimacy and charismatic transformation. But it is Weber's linear theory that he is most remembered for. Weber saw the development of culture as a process of constantly increasing rationalization, of growing inner consistency and coherence. This is most evident in the transition from magic to science; the development of religion from polytheism to monotheism is also viewed in this light (Etzioni & Etzioni-Halevy 1973, p. 5).

Thus for Weber there are both cyclical and linear elements. There is progress at one level but there is a cycle of charisma-routinization-charisma as well.

But it is not only Weber who has traces of both; Adam Smith also includes cyclical elements within his linear march of economic stages. For Smith there are four stages: the era of hunting; the era of pasturage and herding; the era of agricultural; and the era of commercial and exchange economy. The force that moves society is self-love and the love of others. When these combine a society rises, when they are absent a society declines. Smith, aware of Gibbon's "rise and fall" theoretical framework, argued that nations do not necessarily rise; there is stagnation as well.

However, as with Weber, the deeper pattern is linear – the stages are clearly progressive and there is no return to a time of less bureaucracy or back to the age of hunters. Rather within the overall framework and at another level of analysis, it is leadership for Weber and the nation for Smith, that rises and falls. Thus, while they attempt to combine the two, it is a combination of two different levels not an overall theoretical synthesis as in the case of Sorokin.

Merely remaining in the cyclical view and not accepting any theory of progress leads one to a pessimistic view of the future and justifies, as in the case of the theories of Pareto and Mosca, elite rule. If history is but a circulation of elites, a rotation of power from one type of elite to another type of elite, then any revolutionary movement, whether democratic or communist, is a sham. Power will centralize irrespective of who gains it. Democracy legitimizes rule of one class; the military another.

For twentieth century writers, Pareto and Mosca, the cycle operates at the level of governance. Each new elite makes a claim to represent the people, however, it privileges itself or its own class. What hope does this analysis have for efforts to increase participation in governance, to increase public support? Pareto and Mosca, however, would argue that this the wrong question. For Pareto, the circulation of elites shows that for the realization of the good society elite rotation must be regular and often. For Mosca, there should be a balance of the various social forces, the different powers of society military, peasants, landlords, priests, bureaucrats and so forth.

Cyclical views of history privilege structure over human agency. In contrast, revolutionary movements promise a break of structure, an escape from history. It is this rupture that leads to individual dedication. The practical implications of grand theories which relocate individual action to determinism is that they lead to a politics of cynicism. Thus the usefulness of theoretical approaches which attempt to acknowledge the cyclical, the linear, and the transcendental.

5 Metaphors of time

From the view of futures studies, it is the contribution of macrohistory to the study of society-through-time that is of great use. Within macrohistory, many metaphors of time are used. There is the million year time of the cosmos which is useful for spiritual theory but not for social macrohistory. There is individual timelessness or spiritual time, useful for mental peace but not for social development. There is also the classic degeneration of time model from heaven to hell, from the golden to the iron (From *Satya* to *Treta* to *Dvapara* to *Kali* in classical Vedic thought). There is the Chinese model wherein time is correlated with the stars, which thus has no beginning and no end. There is Occidental time which traditionally started with the birth or some other event related to the life of the Prophet. It now relates to the birth of the nation-state. There is also archaeological time, used by Eisler, which is midway between astronomical time (billions of years) and social time (the last few thousand).

In contrast to the linear model and the four stages model which implicitly use the metaphor of the seasons, there is the biological and sexual model.

The rise and fall of nations, dynasties and families can be related to the rise and fall of the phallus. The phallic movement is dramatic and has a clear beginning and a clear end. However, men, it can be argued, (using the linear model) prefer the first part of the cycle imagining a utopia where the phallus never declines. The empirical data suggests, however, that endless rise does not occur.

In contrast, not as obvious to men (and those involved in statecraft and historiography), the female experience is wavelike with multiple motions. Time slows and expands. Instead of a rise and fall model, what emerges is an expansion/contraction model. Galtung, for instance, uses the expansion/contraction metaphor to describe Western cosmology. He also suggests that there might be a relationship between different cosmologies (for example, as Christian cosmology declines, Islamic cosmology might expand).

Expansion/contraction also reminds us that there are benefits in each phase of the cycle. In the contraction, for example, the poor do not suffer as proportionally as the rich who have less speculative wealth available (although certainly the wealthy attempt to squeeze the middle class and the poor as much as possible, especially the poor in the periphery). The expansion/contraction metaphor is also used by Kondratieff and Wallerstein, but for them key variables in the model are prices and the flow of goods, not individuals or social organisms.

Biological time can also be used to understand the future. Ibn Khaldun uses the idea of generational time, of unity and creativity declining over four generations. For Sarkar each collective psychology has its own dominant temporal frame. The *shudra* – worker – lives in the present; the *ksattriya* – warrior – thinks of time as space to conquer; the *vipra* – intellectual/priest – theorises time and imagines transcendental time; while the *vaeshya* – merchant – commodifies time.

But the central metaphor used by all cyclical theorists is the lifecycle. Spengler, in particular, uses this perspective arguing that each individual culture has a unique personality with various distinguishing characteristics. But the cycle has a downward spiral. First there is the stage of culture. This stage eventually degenerates into mass civilization wherein the force of the money spirit leads to imperialism and the eventual death of the culture. For Toynbee, too, civilizations have particular cycles they must go through. Some elites respond to challenges through their creative faculties and others do not meet these challenges. The former expand mentally while the latter intellectually decline. Civilizations that meet challenges expand in size and wealth. Those that do not meet internal or external challenges slowly decline (unless there is rejuvenation from within as Ibn Khaldun argues).

Within the unit of episteme there can be different types of time. The modern episteme, for example, is particularly strong on quantitative, scientific, linear time but weak on mythological, spiritual and seasonal time. Sorokin's stages

also exhibit different models of time. The ideational era, for example, is strong on transcendental time while the sensate is strong on quantitative time.
The best or most complete macrohistory or history of the future must be able to negotiate the many types of time: seasonal, rise and fall, dramatic, mythological, expansion/contraction, cosmic, linear, social-cyclical as well as the intervention of the timeless in the world of time. These must be associated with notions of social structure, individual and transcendental agency.
In general, the ancient cycle alone leads to fatalism and the linear pattern alone leads to imperialism wherein particular collectivities can be placed along the ladder of economic success. Transcendental time alone leads to a focus on the cosmos and neglect of economic progress and social development. For empowering macrohistory, all three are needed. As in models of the future, utopias or plausible futures, few macrohistorians manage to include all these characteristics; rather, macrohistorians privilege certain types of time and avoid or marginalize others. Developing a theory of history or a theory of the future that coherently integrates the many types of time alluded to above is not any easy task and certainly the challenge of the future.

6 The future from macrohistory

We now present the contributions of selected macrohistorians to the study of the future. To do this, we take selected macrohistorians and summarize the key variables they use to think about the future. We will attempt to recontextualize their thought in light of the categories of the modern world (nation-state system, world capitalist system, structures of core and periphery, information and technological revolutions).
This task can be initially be divided into linear and cyclical categories. From Ibn Khaldun we can use three ideas: *asabiya* (unity gained through collective struggle), the rise and fall of dynasties, and the theory of four generations (from creativity to imitation to blind following to indolence). Our questions then become: who are the new Bedouins? Which collectivities are building unity are ready to sacrifice the present for the future? Which ones have struggled a great deal and still retain the warrior spirit? How long will they stay in power? One answer to this question is that the new Bedouins are Japan and the tigers. The Confucian culture provides the unity and hierarchical structure, and defeat in war provides the struggle. How they deal with the current financial crisis will deal us a great deal about the next century.

Figure 1 Ibn Khaldun's model (Daffara, 2004).[4]

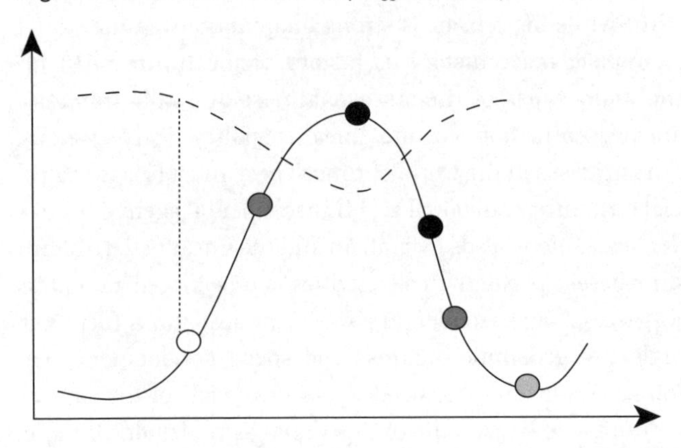

But moving away from the nation-state analysis, it is the social movements who could be the new leaders: the environmental movements, the women's movements and the various spiritual movements. Their unity may develop from struggle against the status quo.

From Sorokin we can use the principle of limits and the pendulum theory of history. What is the next stage in human history? Have we reached the limits of sensate civilization? And if we have, what are the outlines of the emerging ideational or mixed-idealistic civilization?

Figure 2 Sorokin's macrohistory.

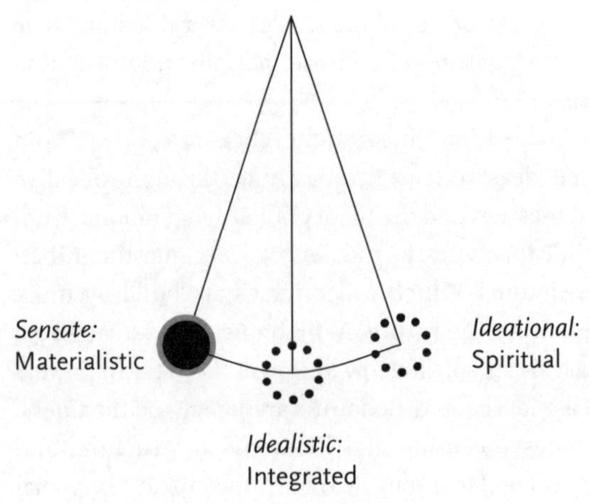

4 These three figures are reproduced with the kind permission of Phillip Daffara. They are from his doctoral dissertation. See: www.futuresense.com.au.

Sorokin also gives us a pattern for the future from which we can understand the formation of the next integrative phase. He places this pattern not at the level of the supersystem but at the level of civilization. Since Western civilization so strongly corresponds with sensate civilization, that is, since the West has assumed the form of the universal system, Sorokin speaks directly to the future of the West. The pattern he gives is *crises, catharsis, charisma and resurrection.* At present, the West stands in the middle of sensate civilization, awaiting the final two stages of charisma and resurrection. The West awaits new leadership that can inspire and lead it to a rebirth in spirit and society, mind and body, individual and collective. But then eventually, since each stage is temporary, the next stage (ideational) will emerge from the integrated stage and the pendulum will continue. But can these categories themselves be transcended? Given the empirical evidence of history and the structure of the real, for Sorokin the answer would be in the negative, at least at the level of the social system. Individually one might adopt a view of the real that is neither ideational, integrated nor sensate, but nihilistic. This latter view, however, does not lead to a social system.

Sarkar is particularly rich as a predictive and interpretive theory of the future (Inayatullah, 1988; 1999; 2002b). From Sarkar, we have his theory of social cycle; his theory of civilization; and, his vision of the future. Appropriate questions to begin an analysis include? Which *varna* will lead next? Which stage are we in now? Will the cycle move forward or will there be a reversal? Which civilizations or ideology will continue and which will collapse or cause oppression? Certainly from the Sarkarian view the former communist (*ksattriyan*) nations are now moving into their *Vipran* era. Will this era be dominated by the church or the university, and how long will it be before these new intellectuals become technocrats for the capitalist era to emerge? Or will there be a slide back to military power? For the nations or groups presently in the capitalist cycle where will the new workers' evolution or revolution come from? And what of the centralization of power that ensues? What will a *ksattriyan* (warrior/military) USA look like? Are we today seeing a battle of *ksattriyan* dimensions of Islam with the *ksattriyan* dimensions of the West? Batra reminds us that historically it is these *ksattriyan* eras that are often seen as the golden ages – at least for those in the centre of the empire – as they provide security and welfare for citizens and expand wealth. Is the emerging world governance/government system the new world ksattriyan system? *Ksattriyan* phases also lead to physically expansion. In the long run, will space be the final frontier?

We can also use Sarkar's theory of civilizations and movements to gauge their possible success. Do these new movements – feminist, ecological, ethnic, regional, and consumer – have the necessary characteristics to create a new system? Do they have an authoritative text, leadership, a theory of political-economy, spiritual practices, fraternal universal outlook, and theory of Being/Consciousness? Are there any ideologies that fulfil these criteria for success? Answering these questions would aid in understanding the long term future of the new movements.

From Toynbee, we can ask which civilizations can meet the numerous technological and ecological survival challenges facing humanity. Which civilizations will find their development arrested as they are unable to deal with the coming challenges? Will there be a spiritual rebirth that revitalizes the present? Is a Universal State next? Or is the next stage a Universal Church? Who and where are the upcoming creative minority? Will Western civilization survive or will it go the way of historical declines? If there is a spiritual rebirth, who will lead it and how will it come about?

Braudel, Kondratieff and other world system theorists offer a theory of economic cycles and war. Are we ready to enter a new expansion or a contraction? If contraction, should we anticipate a global depression? Is war the most likely future since there is no longer a dominant hegemony? What will be the form of the new expansion, and where will it be located? For Kondratieff and other long wave theorists, it is these questions of power and control that derive from the economic that must be asked.

From Ssu-Ma Ch'ien the economic is not an important variable; rather, questions of leadership and the balance of nature are. For example, who will be the sage leader that will return the Tao and restore balance in China-West relations? In Russia, is Yeltsin or his successor the new sage-king that ends the tyrannical dynasty or merely a short term revival in the longer term degeneration? Who is the sage-leader (king) that will provide the similar restructuring for North-South relations? Which nations have moved away from virtue and are now ruled by tyrants? Is the world system moving according to Tao or are there other forces at work? Can government and learning be restored so that there is social balance? How can unity among schools of thought, in the nation and in the family become the dominant trend? As important, how can we reorder our understanding of history and future so to more accurately to reflect the lessons of virtue and morality?

From Spengler the critical variable or tool for understanding the future is the lifecycle of culture. Following Spengler we would attempt to locate cultures in the pattern of the lifecycle. We would ask which cultures are in the final days and which cultures are renewing themselves through interaction with other cultures? We could also ask which cultures are rising and which new cultures are emerging? For example, is Islamic culture in its final stages because of the new religiosity, or is it still expanding because of the recent emergence of the money spirit? Indeed, world fundamentalism could be seen from a Spenglarian view as the last breath of dying cultures. Given that great souls create new cultures, we can survey the world landscape and speculate which thinkers/activists/leaders might potentially create a new culture.

For Vico the next stage in history can be delayed or even eliminated. The final stage of barbarism can be avoided by a warrior-king, a mythic hero. There are also breakouts or disjunctions in these stages. As with other macrohistorians, we should not attempt to develop empirical indicators for each era, stage of gods, heroes, men and barbarism; rather, the effort is insight and interpretation. But we can ask what stage we might be in: are we in the age of barbarians now? Is there a spiritual (Christian) revitalization that can move us out of the cycle? Is the age of the Gods next?

To Pareto and Mosca the theory of elites is paramount. What will be the level of elite circulation in the future? Rapid or fixed? Representations of democracy and widespread participation, notwithstanding, who are the real functioning elites? Who will the future elites be? Is elite rule the only possible governance design? Also of importance is Pareto's different types of elites: the innovators and consolidators. With respect to Mosca, we can ask whether we are moving from a society of the wealthy, to a society of warriors.

From Comte we can ask have we reached the end of the Positive stage? Or, since only a few nations have completely entered the Positive stage, is there still a long wait until the rest of the world joins in and become developed? Or, does the collapse of communism and decline of Islam (in political power if not in mass numbers) signify the continued movement of positivism? Indeed, the present can be construed as a validation of Comte and Smith, among others. Liberalism has become the dominant ideology; the scientific worldview remains the official global ideology.

From Hegel we search for the location of the *Geist*. Which society has solved basic, historical contradictions? In the 1980s, some argued that the *Geist* has shifted from the US to Japan as perhaps the Japanese conquered the contradictions of individual and family in the form of their state? Certainly, it is China and India that hold the world spirit now. But who will the new world historical leaders be? And if we follow Hegel's conclusions, should not we see the ultimate resolution of the *Geist* in the form of a world state either through the victory of one state or through some type of consolidation? In the Hegelian

view, the variables that we should focus on are the dialectics of the spirit, the power of the state, and rare world leaders.
From Marx (with renewal from Wallerstein) we can ask has the end of communism mainly furthered commodification of the world (the proletarization of Eastern Europe)? Will the dramatic and total success of capitalism and its eventual transformation lead to socialism? Are we closer to global socialism than ever before? Will the new electronic and genetic technologies change social relations, or will they merely further commodify workers?
From Adam Smith it is not only the future of the market as a hegemonic metaphor and a site of economic exchange that we should look for but Smith's other key category as well: that of love for the other and love for self as the causal mechanism of social change. Will the future see a society that combines love or self-love or will this combination fail to emerge and lead to civilizational decline?
Spencer's theory and his biological metaphor predicts a world government which would function as the brain of civilization. This world government would also end the rebarbarization of civilization (the world wars). Spencer also predicts a new societal stage, neither barbarous, militant nor industrial. He writes: But civilization does not end with the industrial. A possible future type might emerge.

> *Different as much from the industrial as this does from the militant - a type which, having a sustaining system more fully developed than any we know at present, will use the products of industry neither for maintaining a militant organization not exclusively for material aggrandizement; but will devote them to the carrying on of higher activities (Spencer 1971, 169).*

But Spencer does not detail the contours of this new society. He merely writes that:

> *As the contrast between the militant and the industrial types is indicated by inversion of the belief that individuals exist for the benefit of the State into the belief that the State exists for individuals; so the contrast between the industrial type and the type likely to be evolved from it is indicated by inversion of the belief that life is for work into the belief that work is for life (Spencer 1971, 169).*

In this vision it would be the individual businessman that would lead society onwards. According to economist Robert Nelson, "in social Darwinism, the successful businessman was among the chosen, now the central agent in the evolutionary progress of mankind. Herbert Spencer believed that the end result of progress would be a world without government, marked by altruism in individual behavior" (Nelson 1991, p. 106).

From Eisler the relevant questions relate to gender. What might the partnership society look like? What are its contours and contradictions? How will it come about? What are the supporting trends? What of the contradictory trends which show increased androgyny throughout the planet? Will the partnership society then revert to a cyclical or pendulum social formation or will it continue unabated through the future?
Eisler argues that we hunger for stories and she calls for the new Eves and Adams. By calling attention to ancient Western goddess myths, the Gaia hypothesis, as well as the softer partnership dimensions in all the world's religions, human agency can help create the new story. Eisler gives us many examples of individuals telling a new story, but her main argument is, echoing Kenneth Boulding, if it exists, it can be (Boulding, pers. comm., 9 July 1996). That is, if there are examples of partnership societies either now or in history, we can create a global civilization based on such ethics and values. If it has existed, it can be. By returning to history, she reminds us that such cultures did exist. By foraging through the present and history, she tells us what went wrong, how our pedagogy, our daily actions, our children's stories, our scholarship, our theories all reaffirm the dominator myth. By envisioning an alternative future she intends to create what can be.
Polak focuses specifically on the image of the future. Those collectivities with no vision of the future decline: those with a positive image of the future – transcendental and immanent – advance. Humanity especially now needs a positive image of the future so as to create a new tomorrow. For Boulding, given the power of human agency, the future cannot be forecasted. The image of the future cannot be predicted. As with cultural historian William Irwin Thompson, the image emerges organically at an unconscious mythological level. Mythology cannot be categorized nor rationally created – it is constantly changing always more than what we can know. But although the future cannot be predicted we can assert that history follows a rise and fall related to the image of the future.
We can also ask: why do some societies develop compelling images of the future and others do not. Answering this question would lead to a more complete theory of history. Like Eisler, Boulding's view of the future leads her to develop political strategies in which associations attempt to imagine and commit to their preferred future. A central part of this imagination is faith in the realization of the preferred future. To develop this faith – a concrete belief in a future possibility – Boulding advocates developing future histories in which individuals after imagining their vision develop strategies for how this

vision came to be. From these timelines, hope that tomorrow can be changed is gained. Agency thus overcomes structure.

Sarkar advocates global *samaj* (society, people) movements that challenge nationalism, capitalism and the dogma of traditional religions. Locally and globally active, these movements, Sarkar believes, will transform the inequities of the current world capitalist system. Coupled with spiritual leadership, Sarkar is hopeful that a new phase in human history can begin.

In contrast to image-based macrohistories and the myth based on Gaia is the physics based planetary macrohistory of Nikolai Kardashev and Michio Kaku (Kaku 1998; Kardashev 1964). Their macrohistory is based on the Laws of Thermodynamics and utilization of energy. For them, there are only three energy possibilities: (1) Planet; (2) Star; and (3) Galaxy. Each energy system creates a different type of civilization. There are three types, 1, 2 and 3, with Earth representing Type 0, a civilization attempting to make its way toward Type 1.

- A factor of ten billion separates each civilization, taking up to 200 years to achieve Type 1, 1000 for Type 2 (at 3-5 yearly growth rate) and 10,000 years to achieve Type 3.
- Type 1 Civilization has mastered all forms of terrestrial energy. It can modify planetary weather patterns. The energy needs are so grand that national, religious, sectarian struggles have disappeared or there has been planetary destruction.
- Type 2 Civilization has mastered stellar energy. The energy needs are so grand that they must use the sun to drive their machines (giant spheres, spaceships to channel solar energy to Earth). Growth is managed through the exploration and colonization of nearby star systems.
- Type 3 Civilization obtains energy by harnessing collections of star systems throughout the galaxy. They have already exhausted the energy of their own star. Nothing can destroy this civilization – they can move to other planets, destroy asteroids.

The real danger is in moving from Type 0 to Type 1. The likely scenario is nuclear or ecological destruction. Most likely, our galaxy is strewn with failed civilizations. Type 2 civilizations are looking for ways to contact us but use methods we are unfamiliar with, possibly, and Type 3 could care less, as we are ants or they are in virtual worlds.

The utility in this formulation is that it is logically derived from the laws of physics, it provides us with a clear trajectory, and there are clear choices to be made. Either humanity solves its social problems – nation-states, the great

divide caused by capitalism, religious dogmas – and moves forward creating a planetary civilization, or it self-destructs.
Certainly the way out is a new type of leadership, focused on the needs of the planet as a whole and not on any particular group. This approach to move to a jump in governance can also be applied to organizations. Often, organizations have reached a particular point in their evolution. At this stage, what is needed is jump to a new level, similar to the move from Type 0 to Type 1 civilization. More coordination, integration and complexity are required. And a new type of leader is needed to make this jump. Otherwise, self-destruction is likely.
These macrohistorians aid in transforming the discourse away from the litany of minor trends and events to a macro level of stages and grand causes. While their stages do not provide concrete data for policy making, they provide an alternative way of thinking about the future. The stages also give the study of the future an anchor, a structure from which debate or dialog becomes possible. Otherwise thinking about the future remains idiosyncratic, overly values based.

7 Conclusion

Now, are there any final conclusions that we can make about macrohistory and the future – the social shape of space/time/perspective? First, one central variable in understanding the rise and fall of collectivities, is *creativity*. This is counterpoised to imitation. Creativity leads to expansion, growth, more wealth, more power or inner development. Creativity comes from challenge and is gained through experience. External factors such as resources, geography, invasions are important to macrohistorians but they are not central.
Second is the structure of the stages of history – the *double dialectic*. William Irwin Thompson in his review of macrohistory says it like this:

> *The model of four seems to be a persistent one; it recalls the rule of four in the Indian caste system, Plato, Vico, Blake, Marx, Yeats, Jung and McLuhan. So many people look out at reality and come up with a four-part structure that one cannot help but think that it expresses the nature of reality and/or the Kantian apriori pure categories of understanding. But whether the structure exists in reality or is simply a project of the categories of the human mind is, of course, the traditionally unanswerable question of science (Thompson 1971, p. 78).*

Sorokin has his three stages but there is a fourth stage, a kind of chaotic stage where reality is not fixed at any particular point. This is Sarkar's *Shudra* era,

Galtung's notion of plastic time, or Foucault's postmodern world. Steiner argues that we need a balanced society with three autonomous spheres: the economic, the cultural and the political. All these three interact with the fourth, the environment. In any case, the notion of stages is critical. History must be placed into categories which while simplifying the real at the same time give us more information about a particular age than a mere summation of the particular events of the time.

These stages also directly relate to various metaphysical causes or positions (contradictions or *dilemmas*): mind/body; good/evil; internal/external; expansion/contraction; accumulation/distribution; absolute/relative; theory/data, to mention a few. Any theory of past or of future must confront these basic dilemmas. The weakness in many studies of the future is that these are brushed aside as issues not relevant for tomorrow's society.

The contradiction between city and pasture too must be explained; between the civilized and the barbaric; the courageous and the weak; and the unified and fragmented. For Khaldun it was clear that the urban world brought about the end of *asabiya*. Indeed this concept is central for Spengler's theory of history. Urbanization led to the decline of culture and to mass civilization for Spengler. For others such as Durkheim and Smith, it was the urban world that brought about progress, reduced disease led to economic growth, rationalized the irrationality of pagan religions. The city aided in the development of democratic society and helped move away from traditional society.

However, there must be ways to link the stages. How does each one emerge? The pattern used most often is the *dialectic*. Each stage emerges naturally out of the previous because of the internal contradictions in the previous stage. Sorokin has his principle of limits, Marx and Hegel their dialects, Ssu-Ma Ch'ien has his *yin* and *yang*. The theory of linking historical stages then becomes one of the most important tools in understanding what the future stage might be. The link gives us insight into how the present can or will transform into the future.

When a macrohistory posits a new society which is to be created (as with Marx and Sarkar), then we have the necessity of a vanguard. These are always the minority as Pareto and Mosca have pointed out. They have special access to the real whether because of transcendental reasons, reasons of struggle, or because the system in itself creates their possibility. They are Hegel's world historical leaders, Sarkar's *sadvipras* (balanced leaders), Gramsci's organic intellectuals, Khaldun's bedouins or Toynbee's creative minority.

Figure 4 Sarkar's leaders – sadvipras – at the centre of the social cycle

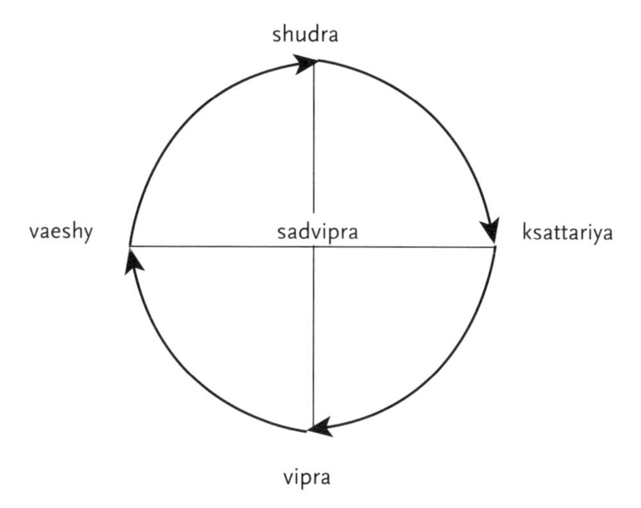

The link between leadership and historical structure is crucial to understanding the possibilities of the future, of the plausibility of creating a different society. For Eisler, Sarkar, Marx, and Gramsci, leadership can transform historical structure. For others such as Khaldun and Ssu-Ma Ch'ien, even as leaders create the future they are bounded by the structures of history, of the rise and fall of virtue, *asabiya*, of the pendulum swings of materialism and idealism. For Hegel, leaders appear to have agency but in fact are used by the cunning of Reason. Leaders merely continue the onward march of the spirit. But for Toynbee, leadership in the form of the creativity minority can keep a civilization from decline, moving it from strength to strength. By meeting internal and external challenges, they can avoid becoming a dominant imitative majority. But for others such as Spengler, once culture has degenerated into mass/mob civilization and the money-spirit has become dominant, there is little any leader can do – the lifecycle of the culture cannot be changed, death inevitably follows life.

Unlike futurists, who speak of disjunction, of bifurcation, of technology transforming the grand patterns of history, macrohistorians by using metaphors such as the birth and death of the individual and the natural world remind us of what does not change, what cannot change. They impose limits of what can be created in the future. But macrohistory is not static. Indeed, it is the macrohistorian's theory of change that is often the insight needed to transform self and other.

As with futurists who do not locate their own work with an episteme, macrohistorians often speak from a view outside of history. While leading to a certain arrogance, this also gives the theory a certain legitimacy, a certain empirical finality. Yet, history is spoken of in dramatic terms, as art, poetry, and as prophecy – not in terms of right or wrong, but in terms of creating a mythic distance from the present (Nandy, 1997, p. 45). Without this prophetic dimension, this privileged perspective of past, present and future, there works would be mere academic treatises that reflect upon history but do not recreate it. Like futures studies, macrohistory is intended to recreate history and future. For the futurist who engages in macrohistory, the key is to be eclectic: to use macrohistory but not be used by it.

References

Daffara, P. 2002, 'Discussion Paper No. 1, Sunshine Coast Habitat Scenarios, 2025-2100', Maroochy Shore Council, Nambour, QLD.

Daffara, P. 2004, 'Sustainable city futures' in *The Causal Layered Analysis Reader: Theory and Case Studies of an Integrative and Transformative Methodology*, ed. S. Inayatullah, Tamkang University Press, Tamsui, Taiwan, pp. 424-438.

Eisler, R. 1995, 'Cultural shifts and technological phase changes: The patterns of history, the subtext of gender, and the choices for our future', Research paper, Center for Partnership Studies, Pacific Grove, CA.

Eisler, R. 1997, 'Dominator and partnership shifts' in *Macrohistory and Macrohistorians*, ed. J. Galtung & S. Inayatullah, Praeger, Westport, CT, pp. 141-151.

Etzioni, A. & Etzioni-Halevy, E. (eds) 1973, *Social Change*, Basic Books, New York.

Faj, A. 1987, 'Vico's Basic Law of History in *Finnegans Wake*' in *Vico and Joyce*, ed. D. P. Verene, State University of New York Press, New York, pp. 22-23.

Galtung, J. & Inayatullah, S. (eds) 1997, *Macrohistory and Macrohistorians*, Praeger, Westport, CT, London.

Hayward, P. & Voros, J. 2006, 'Playing the Neohumanist Game' in *Neohumanist Educational Futures: Liberating the Pedagogical Intellect*, eds. S. Inayatullah, M. Bussey, I. Milojevic, Tamsui, Tamkang University, pp. 283-296.

Inayatullah, S. 1988, 'Sarkar's spiritual-dialectics: An unconventional view of the future', *Futures*, vol. 20, no. 1, pp. 54-65.

Inayatullah, S. 1999, *Situating Sarkar*, Gurukul Publications, Maleny, QLD.

Inayatullah, S. 2002a, *Questioning the Future: Futures studies, action learning and organizational transformation*, Tamsui, Taiwan, Tamkang University Press.

Inayatullah, S. 2002b, *Understanding Sarkar*, Brill, Leiden.

Inayatullah, S. (ed.) 2004, Special issue on Macrohistory, *Journal of Futures Studies* (Contributing authors include: Anthony Judge, Phillip Daffara, James Dator, Yongseok Seo, Jay Weinstein, Walter Truett Anderson, William Halal, Anodea Judith, and Jay Earley), vol. 9, no. 1.

Kaku, M. 1998, *Visions: How Science Will Revolutionize the 21st Century and Beyond*, Oxford University Press, Oxford.

Kardashev, N. 1964, 'Tranmission of information by extraterrestrial civilizations', *Soviet Astronomy A7*, vol. 8, pp. 217-221.

Nandy, A. 1997, 'The futures of dissent', *Seminar*, no. 460, pp. 42-45.

Nelson, R. 1991, 'Why capitalism hasn't won yet', *Forbes*, November 25.

Spencer, H. 1971, *Structure, Function and Evolution*, Michael Joseph, London.

Thompson, W.I. 1971, *At the Edge of History*, Harper and Row, New York.
Voros, J. 2006, 'Nesting social-analytical perspectives: An approach to macro-social analysis', *Journal of Futures Studies*, vol. 11, no. 1, pp. 1-22.

7 Geography, planning and the future

Ela Krawczyk

1 Introduction

Geography, by its definition, is concerned with space. It seeks to describe and analyse the spatial distribution of physical, biological, and human phenomena that occur on the surface of the planet, and to examine their interrelationships (Encyclopaedia Britannica 2007). Time, although it seems to be a logical variable for describing change in spatial conditions, for many years tended to be regarded by geographers as "not its province" (Crang 2005). The intrinsic relationship between time and space was finally acknowledged and examined in the 1960s by a Swedish geographer, Torsten Hägerstrand, whose work led to the development of a new concept – Time Geography. The notion of time is usually discussed in the broad geographical literature from two perspectives – of the inherent link between time and space as represented in Time Geography, or a philosophical and social concept that is being introduced to geographers, rather that being a specific part of geographical thought. The future, like past and present, is perceived as a notion within a broader view of time.

The relationship between geography and the future is best represented through the discipline of 'spatial planning'. As the 'spatial' dimension forms geography's heart, 'planning' is an activity that is inherently concerned with the future. Although, contemporary geography and spatial planning form two distinct disciplines, their spheres of knowledge overlap significantly. Looking back, it could be argued that geography was one of the primary scientific disciplines that gave birth to spatial planning. Often, browsing the shelf with geographical and spatial planning literature it is difficult to clearly see which publications belong to which discipline, especially when urban geography and urban planning are concerned.

The main aim of this book is to explore what role the term 'future' has played and continues to play in different scientific disciplines. This chapter attempts to consider the position of 'future' in geography and the spatial and urban planning which are a part of the 'spatial sciences'. As the first step, I will try to examine the relationship between time and space and to develop an understanding in what way geographers think about time in general. Recognising that there is a very little attention given specifically to the concept of future in the broad geographical thought, but the link between spatial dimension and the future is most clearly visible in spatial and urban planning, I will review the evolution of the planning's approach towards future over time, focusing in the final section on the present role of futures thinking and approaches in contemporary planning.

2 Geography: space and time

Many geographers see time as the fourth dimension of space (Belbin 2002; Crang 2005); however, their perception of the relationship between space and time varies. Belbin (2002) believes that space is not just a 'spectator' but a 'player' that even dominates time. "What is a year but movement in space: one orbit of the Sun by the Earth? What is a day but movement in space: one spin of the Earth on its axis of rotation? If these spaces change, then time changes (*ibid*: p. 1-2)." This view reinforces the dominance of space over the time. However, geographers are not interested only in the static patterns or identifying location, but they focus on measuring and assessing dynamic patterns of flow, variability, spatial diffusion of matter, energy and ideas over time. Therefore, time is one of the key axes for the description of the changes taking place in space. Crang (2005, p. 200) sees space as "neither self-evident nor self-sufficient, but rather often mutually and problematically defined by and with problematic concepts of time." He argues that although many geographers see space and time as obvious categories that not require further scrutiny, both notions have multiple facets and definitions and both concepts as well as their relationship should be explored more deeply. In his view geographers should develop a broader understanding of time as a philosophical and social concept, and not only as a notion interlinked with space.

Torsten Hägerstrand, the father of Time Geography, examined space and time within a general equilibrium framework, in which it is assumed that every entity performs numerous roles; it is also implicitly admitted that location in space cannot effectively be separated from the flow of time. In this framework,

an entity pursues a space-time path, starting at the point of birth and finishing at the point of death. Such a path can be represented in space and time by forming both spatial and temporal dimensions into a space-time path. Time and space are recognised as being inseparable (Wachowicz 1999).
Time Geography has provided a foundation for detecting entities space-time paths and for developing of an understanding of potential spatio-temporal relationships between them. Furthermore, the application of time geography has led to the development of the concept of a 'continuous path' that is used to represent the experience gathered during the life-span of an entity. This experience is in fact conceptualised as a succession of changes of locations and events over a space-time path. Time Geography has been mainly applied in modelling individual activity paths within a period of time, including analysis of the pattern of activities as well as simulation of individual activity paths (*ibid*).
Although, there is not much attention given to the concept of time and the future in broad geographical literature, geographers do recognise the role they can play in influencing the spatial future. Belbin (2002, p. 2) states:

> *"Geographers can advise on how spatial patterns can be improved and predicted: the best location, the best use of an area, the spatial targeting of mitigation measures: the best practice given prevailing conditions and limitations, the most likely future change to a boundary, the most likely change of use for an area, the most likely spatial impact of a certain event, the most probable spatial scenario given a set of initial conditions and so on."*

They focus on predictions related to space activity and finding a highly probable, optimal use for the area. Thus, geographers mainly use modelling, GIS, and various prediction and projection techniques in approaching the future.

3 Planning and the future

Planning can be defined as "the making of an orderly sequence of actions that will lead to the achievement of a stated goal or goals" (Hall 1992, p. 3), "a process of human forethought and action based upon that thought" (Chadwick 1971, p. 24) or as a method for the implementation of a selected future (Serra 2001). Definitions of planning link the present activities with the future emphasising the inherent relationship of planning with the future. Myers (2001, p. 365) captures it adequately: "The future is generally considered to be a core concern of the planning profession."

In this chapter I will focus on the concept of future in the context of urban planning, as I am particularly interested in this topic. Overall, urban planning can be described as an activity pursued in an attempt to achieve certain social and economic goals, in particular to shape and improve urban environments, where increasing numbers of people live (Encyclopaedia Britannica 2005). However there are many different interpretations of what planning actually constitutes. For example, Hall refers to 'urban planning' as:

> *"... the planning with a spatial, or geographical, component, in which the general objective is to provide for a spatial structure of activities (or of land uses) which in some way is better than the pattern existing without planning" (Hall 1992, p. 4)*

Sharp (1940) defines planning as an attempt to formulate the principles that should guide human actions in creating a civilised physical environment for people. According to Ratcliffe (1981, p. 13) planning should provide "the right site, at the right time, in the right place, for the right people" and according to The European Consultative Forum on Environment and Sustainable Development (ECFESD) (1999) spatial planning should "balance public interests between, on the one hand, the objectives of social cohesion and sustainability and, on the other, the need of competitiveness and market imperatives ECFESD 1999, p. 4). Rose describes urban planning as "a multi-dimensional activity that ought to be integrative, embracing social, economic, political, psychological, anthropological and technological factors" (Rose 1974, p. 26). He argues that the future-orientation of planning arises from its 'intrinsic' purpose of achieving human 'betterment' and 'improvement'.

Finally, Bracken (1981) offers interpretation of planning in terms of its approach towards the future. He recognises four modes of planning: planning for the present by reacting to past problems; planning towards a predicted future; planning with a predicted future; and planning by creating the desired future. *Planning for the present by reacting to past problems* involves problem analysis, designing interventions and allocating resources accordingly. *Planning towards a predicted future* focuses on determining trends and deciding whether they are positive and negative. Resources are allocated in line with the desires to promote or change these trends. *Planning with a predicted future* also involves identification of trends, but the resources are distributed in a way that enables the use of these trends to the advantage. Finally, *planning by creating the desired future* requires deciding on which future is desired. Resources are allocated in order to change existing trends or create new ones. The desired future can be based on the existing or predicted trends, or new values.

Having explored the link to the future in different views of urban planning, it's time to examine how future was approached and constructed in the urban planning processes over time. Using a broader context of the evolution of the planning field, I will try to examine the following issues: the way of thinking about the future; attitudes towards change and complexity; the way of thinking about the city and its various dimensions; methods and techniques used for the exploration of the future; people and institutions involved in the planning process; and the weaknesses of the planning process in relation to the future. The summary of the evolution of planning's approach towards the future is presented in Table 1. The tackling of these issues can vary between countries, as urban planning has been practised differently by different nations. Although, I have tried to keep the broadest perspective possible, the main sources used refer to the British and Irish planning practice.

3.1 *Urban planning until 1945*

The beginnings of the modern planning thought can be traced to the period of the late 19th and early 20th century. This period is well summarised by Wachs (2001:368): "idealised futures and grand visions were guideposts for current actions". The need for planning was triggered by the simultaneous and related processes of industrialisation and urbanisation and their impact on cities. At first, urban environments were struggling with two main issues: provision of housing and health problems. Subsequently, World War I and the economic crisis during the 1930s brought about a number of social and economic problems – unemployment, poverty and the dereliction of urban areas (Rydin 1998).

Planning, prior to the 1st World War, was performed mostly by architects and urban engineers. After the war the 'planning profession' started to gain recognition, and increasingly 'planners' were charged with planning. Planning activity was based on the assumption that there is a relationship between physical environment and society (physical determinism). It was believed that through the careful professional design of urban territories, it was possible to create an environment that would enhance the quality of life and also would improve inhabitants themselves – physically, morally and socially. Therefore, the main focus was placed on the physical aspects of urban design (*ibid*).

Table 1 The evolution of the approach towards future in urban planning

	Up to 1945	1945-mid 1960s	Mid 1960s and 1970s	1980s	1990s – present
The way of thinking about future	Visions of the future -physical states of cities enclosed in blueprint plans	Future as a fixed known end presented in a form of master plans	1. Recognition that future can unfold differently 2. Exploration and evaluation of possible courses of action	1. Future orientation of planning pushed away 2. Visions of future created with participation of citizens (USA, Canada)	1. Renewed interests in future-orientation of planning 2. Future seen as uncertain, with many possible alternatives
Attitudes towards change and complexity	1. Little or no attention given to change 2. P. Geddes pointed out that planning should take the processes of change into account	1. Control and regulation of the pace and direction of change 2. Lack of recognition for complexity	Recognition of complex and dynamic nature of cities, as well as the uncertainty of the future	Complex and continuously changing nature of urban territories	1. Recognition of instability, rupture, discontinuity and accelerating pace of change 2. Complexity of cities and the world within they operate 3. High degree of uncertainty
The way of thinking about the city and its various dimensions.	Focus on the physical aspect of cities underlined by 'physical' determinism	1. Main focus on the physical planning and design of land use and built form 2. No relation to economy, apolitical	Cities treated as systems including physical, social, economic and political dimensions	Cities as systems	City as a comprehensive whole
Methods and techniques used for the exploration of the future.	1. Personal visions of the future prepared by architects or engineers 2. P. Geddes's approach: 'survey-analysis-plan'	1. Planning based on 'survey-analysis-plan' approach 2. Projections for population and transport	Modelling Forecasting	Forecasting Modelling Visioning First 'swallows' of strategic planning and futures methods	Forecasting, Modelling Strategic planning Visioning SWOT analysis Scenario analysis Various futures approaches and methods

Planners focused mainly on the production of master plans, this is the statements about the future end-state of the city or the region. These reflected their visions of how cities should look. As Hall described it: "There was one true vision of the future world as it ought to be, and each of them saw himself as its prophet" (Hall 1992, p. 61). An interesting example of such vision is the concept of 'garden cities' developed by Ebenezer Howard, and realised earlier by a number of industrialists with 'philanthropic leanings', for example Robert Owen, Titus Salt and George Cadbury. Most of those early plans and designs did not have alternatives, and little or no attention was given to the changes and trends taking place in the outside world. An exception was Patrick Geddes, who believed that planning should start with the world as it is, and it should try to take into account and work with economic and social trends, rather than impose its own subjective vision of the world (*ibid*). Geddes's planning theory was based on the approach: 'survey, analysis, plan'. He argued that the planning process should start with understanding the city within its historical context. This was to be achieved with the survey method. The role of survey was to inform and stimulate public discussion and co-operative action. Geddes also proposed to examine cities in their regional context (Hebbert 1982). Furthermore, he was an advocate of 'visionary thinking and ideas' in shaping urban life (Hasselgren 1982).

In the period before 1945, planners thought about the future as an end-state. The end-state was usually presented in the form of grand visions of idealised future. There was no recognition that many alternative futures are possible and no attention was given to trends and changes taking place in the world, with the exception of Patrick Geddes work. Looking back today, it could be said that it was a very simplistic and undeveloped approach towards the future based on visions of what was desired and a belief that it can be achieved without considering present and future trends and changes.

3.2 Post-war period (1945 - mid 1960s)

The first decades after the war in Europe and Britain were characterised by high population growth and strong economic development, full employment and the increasing migration of people. The planning activity was performed by planners – professionals from mainly architectural, engineering and surveying backgrounds, although increasingly emerging from disciplines such as geography, economics and sociology (Rydin 1998). Planning professionals working for local governments became an important part of an elaborate planning system set up in Britain and many other countries in the post war period. The essential function of the planning system was "to control and regulate the pace and direction of change – social, economic and physical" (Hall 1992:115). It was assumed that the control of change was both desirable and feasible. It was desirable because most decision-makers believed that uncontrolled change had unwanted and negative effects. The feasibility of this approach was based on the assumption that population growth and economic development would be slow; therefore, it was possible to control such change (*ibid*).

In the first years after World War II, similar to the interwar period, urban planning was focused on the physical planning and design of land-use and the built form. The vision of the desired future was presented in a form of master plans or blueprints, which defined the required overall pattern and size of towns and cities across the whole country and detailed spatial structure along with the layout of urban territories or their constituent parts (Taylor 1998). The planning process, through which the master plans were constructed, was based on Patrick Geddes's procedure: 'survey, analysis, plan'. The existing situation was surveyed and then analysed in order to identify the remedial actions that needed to be taken. Then these actions were incorporated into the plans. Such spatial plans were prepared every five years. The sequential character of the planning process created an opportunity to review, update and modify plans, activities considered as crucial in the continuously changing world (Hall 1992).

The relation of master plans to the social dimension of urban environments was based on 'physical determinism'. Planners believed that communities and social neighbourhoods could be created by planning the physical neighbourhoods (Taylor 1998). Physical planning did not have much relation to the economy (Rydin 1998) and was seen as an apolitical process (Cherry 1974). Planning started to be seen as a part of politics only in the 1960s as a result of new political and social science theories (*ibid*).

'Blueprint' or 'master' planning was criticised for a lack of adequate empirical analysis and understanding of how cities and towns actually worked. Planners fixed the future of urban territories in blueprint plans, failing to recognise their changing nature as well as the complexity, richness and difficulties of social existence in cities (Taylor 1998). The plans were unsuited for coping with the broad and long-term policy issues, as well as with regional and sub-regional matters. They were inflexible and unable to adapt swiftly to changing conditions and circumstances (Bruton 1974).

In the period between 1945 and mid 1960s planner's perception of the future have evolved significantly. Although, master plans and blueprints of the desired future were still the main ways of portraying the future and alternative futures were still not considered, it was recognised that the world around is changing and these changes and trends need to be analysed. As the pace and complexity of changes were much lesser than today, planners believed that they were able to control and regulate the pace and direction of change. Future was still seen mainly as the preferred future state and no uncertainty or complexity were attributed to it.

3.3 The mid 1960s and 1970s

Radical changes in planning thought, which took place in the second half of the 1960s, were a result of the strong critique of 'master' planning, developments in social and political sciences and growing importance of cybernetics[1] in various areas of life. McLoughlin (1969, p. 24) postulated:

> *"Whilst retaining the vitally important understanding of the operations of building, engineering and the measurement and transference of land, the profession needs a far greater awareness of the process of change in the human environment, the underlying reasons for them, their manner of accomplishment, the complex web of interactions between human groups and much greater skill in the techniques of foreseeing and guiding change."*

The new thinking was channelled into two directions: the 'rational process' view of planning and the 'systems view' of planning. The 'rational process' view of planning was concentrated on the planning process itself, which was seen as rational process of decision-making, while the 'systems view' of planning was focused on an object (town, city, urban region) that ought to be planned (Taylor 1998).

The 'rational process' regards planning as a method for reaching decisions was advocated by authors such as Webber (1963). It has its origins in corporate thinking and management, where changes in industrial planning thought led to the development of a decision-making science (Hall 1992). This view of planning seeks to improve the decision-making process to enable the making of more rational decisions. A rational decision can be described as "one where all the various alternative courses of action are considered, the consequences resulting from them are identified and compared, and preferred alternatives selected" (Ratcliffe 1981:117).

The 'systems view' of planning, promoted by McLoughlin (1969) and Chadwick (1971), considers cities, towns and urban regions as 'systems', recognising

1 Here the term 'cybernetics' is used in the meaning: "a new way of organising existing knowledge about a very wide range of phenomena" (Hall 1992, p. 229).

their complex nature. McLoughlin (1969) described urban systems as *human activities* that occur within adapted *spaces* (buildings, arenas, parks, lakes, forests and so on) and are linked by *human communications* and *transport* (communications concerned with material interaction – goods and people). He emphasised the dynamic character of systems manifesting itself through continuous change.

The emergence of the 'systems' view of planning as well as the 'rational' view brought a huge shift into the way the future was approached and constructed in the planning processes. The future was no longer a fixed state within master plans, but it was recognised that different courses of action were possible. In other words, 'master' planning could be described as "planning with confidence and certainty towards a known end product" (Cherry 1974:81), and the 'systems' planning as "planning for uncertainty, simply making most effective use of existing resources and resisting dramatic interventions in the system to achieve some desired end (*ibid*)".

The 'systems' view of planning constituted the basis for the development of the cyclical planning process. This process, proposed by McLoughlin (1969), has six basic steps as described below: examination of the environment; formulation of goals; study of possible courses of action; evaluation of possible courses of action; action, and review. An important issue in regard to the identification and evaluation of alternative courses of action were the methods used for that purpose. Among the main techniques employed, both in the 'systems' and 'rational' view of planning, were modelling and forecasting. Models were created to represent the behaviour of the system over time, both in the recent past and in the future. Planners used them to gain an understanding of the impact of alternative courses of action that may unfold (Hall 1992) or, stated differently, "to understand the forces that determine the size and nature of urban areas and the location of land uses therein" (Ratcliffe 1981. p. 267). Bracken (1981) warned against attempts of using them in the generation of plans and policies as that raises methodological and philosophical doubts.

Forecasts, the predictions of future developments generated through application of various forecasting techniques, were recognised as essential inputs to decision-making and planning processes (Ratcliffe 1981, Wheelwright & Makridakis 1985). An 'explosion' of forecasting activity was observed in the 1960s and 1970s (Ratcliffe 2002). The wider availability of computers encouraged their extended use in forecasting and contributed to the dynamic development of quantitative prediction techniques (Wheelwright & Makridakis 1985), such as mathematical models, simulation, system analysis, cost-benefit and cost-effectiveness analysis (Steiss 1974).

The huge popularity of forecasting observed in the 1970s brought a concern that forecasts, like models, can become solutions to problems instead of fulfilling their primary role of supporting and informing decision-makers (Couts 1969; Bracken 1981; Ratcliffe 1981). Couts (1969) emphasised that decision-making should be based on the reliable thought process and analytical thinking, while forecasting and other techniques should enable generation of necessary information and analysis to assist decision-making. Bracken (1981) stressed the necessity to understand the assumptions and information on which forecasts are based, as these assumptions and data continuously change.

The weak side of forecasting is its inability to warn about forthcoming events and changes. Great disappointments arose from considerable prediction errors that led to mistakes in decision-making and planning. The main weaknesses of forecasting that affected planning were: unexpected developments, foreseen events that never took place, errors in timing and the level of predicted changes. The limitations of forecasting are related to its nature, as the prediction process is based on the assumption that the future will be akin to the past and present (Wheelwright & Makridakis 1985).

The first critiques of the 'systems' approach arose in the mid 1970s. The approach was criticised for a number of reasons. The scientific nature of the 'systems' approach suggested that the world could be completely understood and its future states predicted. This, of course, was far from the truth. This approach led to establishing the planner's role as the 'superior scientific expert'. There was a lack of public participation and consultation. And finally, the society was treated as 'homogeneous aggregate'. For instance, it was accepted that the welfare should be maximised, but there was no distinction between different spatial and social circumstances in distributing this welfare (Hall 1992).

The period between mid 1960s and the end of 1970s brought significant changes to the way of thinking about the future. Future was no longer perceived as a fixed state (blueprint), but it was recognised that there are many possible alternative futures. The appreciation of the importance of change in the previous years now turned into strong emphasis being placed on change and its consequences. The images of the future were created through the extrapolation of the existing trends and the exploration of their implications. The ability to foresee and forecast future changes became very important, which led to the development of the whole range of the forecasting techniques. At this point, it is necessary to stress that up to the end of 1980s the exploration of the future was done mostly on the short-term basis.

3.4 *The 1980s*

In the 1980s planning theory in Britain was developing in a number of directions, for example 'implementation' theory, planning as 'communicative action', and 'problem-centred' planning (Taylor 1998). None of these theories brought significant changes to the way the future was approached and constructed but rather quite the opposite. Many authors seem to share an opinion that strategic planning was in 'limbo' (Healey *et al.* 1997; Saler &nd Faludi 2000, Albrechts *et al.* 2003). This situation was an effect of policies of the Thatcher government, which involved the abolition of provincial metropolitan county councils that had hitherto served the functions of strategic planning authorities. These policies also intended a reduction in the role of public planning in favour of private initiatives (Hall 1992). Andrew Isserman (1985) in his essay *Dare to plan* argued that planners have forgotten that the role of planning is to lead from the present to the future. He accused the planning profession of 'abandoning' its primary responsibility to inspire and inform about visions of probable and desirable futures in order to fulfil its problem-solving orientation and to actively participate in short-term activities such as budgeting, public/private land development, funding of social services, programme management and project administration. Isserman (1985) distinguished several causes for the 'neglect' of the future in planning.

- Planning's shift from the architecture and design towards the social sciences and scientific methods.
- Budgets cutbacks and an atmosphere making idealism, visionary thinking and inspiration obsolete.
- The pressures arising from the daily job requirements.
- Planners scepticism and lack of confidence in their ability to think meaningfully about the future and to stimulate change.

Along with 'future neglect', a different approach towards the future was emerging in the United States and Canada, where visions and visioning were applied in urban planning in the early 1980s. Among the first places to use visioning was Tennessee, followed by other regions, including: Quebec (Canada), Washington, Oregon, Iowa, Illinois, Minnesota, Maine, Connecticut, rural New York, Massachusetts, Arkansas and the Carolinas (Shipley 1998). Visioning, "the notion of creating images of the future to serve as goals or guides for planning decisions" (Shipley 2002, p. 7), involves the participation of citizens in imagining and to a lesser degree constructing the future (Puglisi 2000). The emergence of visioning indicated that a key transformation was taking place in two areas – the way of approaching and constructing the future and public participation in planning.

3.5 *The 1990s to present*

The decade of the 1990s and first years of the 2000s can be characterised, among other things, by a growing interest in the future dimension of urban planning. As Myers and Kitsuse (2000) put it:

> *"Recent writers have proclaimed the future orientation of planning as unique to the field's identity and have called for renewed focus and development of future-oriented skills."*

A number of reasons can be listed to explain this shift: challenges posed by the contemporary change; competition between cities and urban regions; sustainability agenda; urban governance; and other.

In order to respond to these challenges, cities and urban regions have been increasingly applying various forms of strategic planning and other future-oriented approaches, for example Foresight and Prospective. Parrad (2004), who reviewed various projects representing futures thinking in European cities, has identified four different types of exercises: 'strategic planning' activities; 'strategic competitiveness' projects in order to increase economic competitiveness of cities and their regions; 'metropolitan projects'; and exercises driven by the agenda of 'sustainable development'.

Strategic planning is one of the most widespread types of futures-oriented activities. It has been used by business and the military since the 1950s (Mintzberg 1994). In the context of urban planning, it can be described as:

> *"... a social process through which local communities respond to internal and external challenges with respect to the management of local environments... local communities build new strategic ideas and policy discourses, build institutional relations, and mobilise political support" (Healey et al. 1997, p. 293).*

Strategic planning aims at the development of a long-term, comprehensive vision for the city or urban region, which comprises all urban elements: physical, economic, and social. It is not a process exclusive to planners, but its success depends on the inclusion of all major stakeholders (Tosics 2003). An evaluation of the relevant literature reveals a vast number of cities and urban regions, as well as whole countries, performing various types of strategic planning. Among the cities that have developed future-oriented exercises are: Munich, Vienna, Budapest, Warsaw, Prague (Tosics 2003), Hanover, Flanders, Northern Ireland (Albrechts *et al.* 2003), Glasgow and the Clyde Valley (Goodstadt 2001), Vancouver, Hong Kong (Freidmann *et al.* 2004), Helsinki, Venice, Utrecht, Birmingham, Brno (Parrad 2004).

Also, over the past two decades, the urban planning profession has increasingly been reaching to the futures field in a search for methodologies and techniques that could be used in urban planning. Cities like Lyon, Barcelona, Bilbao, are being recognised as 'flagship' examples of the successful application of a futures approach in planning (EC 2002; Krawczyk & Ratcliffe 2004). Other cities, such as Lincoln, use futures approaches (Prospective Through Scenarios) but on much smaller scale in comparison to those listed earlier. Many cities also use individual futures techniques, such as scenarios (Edinburgh, Gipuzkoa, Göteborg), visioning (Amstelveen), Delphi survey (Malaga), however, it is often difficult to distinguish whether their overall approach is a 'futures' one.

The futures exercises undertaken by towns, cities and regions vary in regard to their aims, structures, budgets, timescales, time horizons, and methodologies. Most of them are based upon informal approaches and face difficulties arising from their innovativeness (Parrad 2004). A driving force behind these projects is a strong need for gaining control over the future of the territory. This need can arise from a necessity to deal with the existing problems, a requirement to improve economic competitiveness and re-brand, an aspiration to achieve sustainable development, or often a mixture of the above (*ibid*). Development of the preferred future is usually based upon mobilisation and collaboration between all actors and stakeholders within the region and the urban territory is approached in a holistic and integrated manner. It is recognised that the importance of the actual process is equal to the importance of the final product. In many cases the main interests in the projects lies in more indirect results, such as raising the awareness and building social networks (*ibid*).

Over the last 15 years or so, planners' approach towards the future has been changing significantly. For the first time, future started to be considered not only on the short-term but also on the medium- and long-term basis. Future has been moving towards the centre of planning activity as the pace of change speeds up and complexity is growing and, consequently, uncertainty increases. Planners increasingly recognise that it is important not only to explore various future options to prepare for what it may hold, but also to develop visions of preferred future in order to shape the future.

4 Final thoughts

Time, by geographers, is mostly considered to be the fourth dimension of space. In the view of Time Geography, it is inseparable from space dimension of a space-time path pursued by an entity. Geographers rarely talk about time as a separate concept, and even more rarely about the future itself. However, they do recognise that through an understanding of changes taking place in space they can have an important input into improving and predicting spatial patterns in the future. The relationship between geography and the future is best represented through 'spatial planning' and geographers' understanding of spatial patterns and interrelationships can best be utilised by the planning activity.

In this chapter, I have tried to show how the concept of the future have changed and evolved over time in the urban planning context. Starting from the time before Second World War when the future was represented through idealised future images of the physical aspect of cities developed by visionary architects and urban designers, going through the period of 'blueprint' and 'masterplans' up to 1960s, then through the significant shifts in the planning mindsets that resulted in the development in 'systems' and 'rational process' view of planning in the late 1960s, and 'the future in limbo' in the 1980s, to finally arrive at present when a renewed interest in the future orientation of planning is visibly taking place.

Over the last two decades, a new context for the urban development has emerged as a result of the processes of globalisation, the sustainability movement, and changes in the government systems, all of which are underlined by increasing complexity and uncertainty of change. This forced urban planners and decision-makers to look for new improved ways of approaching and constructing the future that would enable them to deal with present problems and prepare effectively for the future. As a result, urban territories have been more and more engaging in various forms of future-oriented activity using strategic planning and various futures methods. Although, clearly, futures methods have been increasingly incorporated into planners' toolkit, I would argue that there is also a strong need to develop a 'futures mindset' amongst planners and decision makers that would change their underlying perceptions about the future and consequently help to use these methods to their full potential. Such a 'futures mindset' would bring the recognition that at any given moment an indefinite number of different futures is possible, the ability to explore what alternative futures are possible and probable, the capacity to test plans and decisions against the alternative futures, the development of preferred future visions and the creation of robust and flexible plans and policies that would reduce the threats and help to exploit opportunities in the future.

References

Albrechts L., Healey P., Kunzmann K.R. 2003, "Strategic spatial planning and regional governance in Europe", *APA Journal*, Vol. 69, No. 2, pp.113-129.

Belbin S. 2002, *The future and geography*, [Online], [Accessed on 19th January 2007]. Available from World Wide Web: http://www.scienceinafrica.co.za/2002/february/geography.htm

Bracken I. 1981, *Urban planning methods: research and policy analysis*, Methuen, London and New York.

Bruton M. ed. 1974, *The Spirit and Purpose of Planning*, Hutchinson, London.

Chadwick G. 1971, *A Systems View of Planning*, Pergamon Press, Oxford.

Cherry G.E. 1974, *The development of planning thought*, in Bruton M. ed., *The Spirit and Purpose of Planning*, Hutchinson, London.

Couts D. 1969, *'What is forecasting?'*, in Yewdall G. ed., *Management Decision Making*, PAN.

Crang M. 2005, *Time:space*, in Cloke P. and Johnston R., *Spaces of geographical thought deconstructing human geography's binaries*, Sage Publications, London.

EC 2002, *Practical Guide to Regional Foresight in Ireland*, Office for Official Publications of the European Communities, Luxembourg.

ECFESD 1999, *The European Spatial Development Perspective (ESDP): Comments and recommendations from the European Consultative Forum on Environment and Sustainable Development*. [Online],[Accessed on 20th April 2005]. Available from World Wide Web: http://europa.eu.int/comm/environment/forum/spatreport_en.pdf.

Encyclopaedia Britannica (2005) *Urban planning*. [Online], [Accessed on 19th February 2005]. Available from World Wide Web: http://www.britannica.com/eb/article?tocId=9074444andquery=urban%20planningandct.

Encyclopaedia Britannica (2007) *Geography*. [Online], [Accessed on 19th February 2007]. Available from World Wide Web: http://www.britannica.com/eb/article-9036464/geography.

Freidmann J., Bryson J., Hyslop J., Balducci A., Wiewel W., Albrechts L. 2004, "Strategic spatial planning and the longer range", *Planning Theory and Practice*, Vol. 5, No. 1, pp. 49-67.

Goodstadt V. 2001, The need for effective strategic planning: the experience of Glasgow and the Clyde Valley, *Planning Theory and Practice*, Vol. 2, Is. 2, pp. 215-221.

Hall P. 1992, *Urban and regional planning*, Routledge, London.

Hasselgren J. 1982, *What is living and what is dead in the work of Patrick Geddes*, in: *Patrick Geddes: A Symposium*. Ocassional Paper in Town and regional Planning, Duncan of Jordanstone College of Art / University of Dundee.

Healey P., Khakee A., Motte A., Needham B. eds. 1997, *Making strategic spatial plans. Innovation in Europe*, UCL Press, London and Bristol.

Hebbert M. 1982, *Retrospective on the outlook tower*, in: *Patrick Geddes: A Symposium*. Ocassional Paper in Town and regional Planning, Duncan of Jordanstone College of Art / University of Dundee.

Isserman A. 1985, "Dare to plan: An Essay on the role of the future in planning practice and education". *Town Planning Review*, Vol. 56, No. 4, pp. 483-491.

Krawczyk E., Ratcliffe J. 2004, *Imagineering cities: crating liveable urban futures in the 21st century*, in: WFS, *Thinking Creatively in Turbulent times*, WFS, Bethesda, Maryland.

McLoughlin J.B. 1969, *Urban and regional planning: a systems approach*, Faber and Faber, London.

Mintzberg H. 1994, *The rise and fall of strategic planning*, The Free Press, New York.

Myers D. 2001, "Symposium: Putting the Future in Planning, Introduction", *Journal of the American Planning Association*, Vol. 67, No. 4, pp.365-367.

Myers D., Kitsuse A. 2000, "Constructing the Future in Planning: a survey of theories and tools", *Journal of Planning Education and Research*, Vol. 19, No. 3, pp. 221-31.

Parrad F. 2004, *Future thinking in European cities: evolution of strategic urban planning across Europe*, Synthesis of a study for the French ministry of urban development.

Puglisi M. 2000, "Futures Studies and the challenges of participatory urban planning: the use of interactive scenarios in Barletta", paper presented at the seminar "The Quest for the Futures", Turku School of Economics and Business Administration, Turku, Finland, June 13-15, 2000.

Ratcliffe J. 1981, *An Introduction to Town and Country Planning*, UCL Press, London.

Ratcliffe J. 2002, "Imagineering Cities: creating future 'Prospectives' for present planning", conference paper presented at Turkish Real Estate Seminar III, 2-4 May 2002, Istanbul.

Rose E.A. 1974, *Philosophy and purpose of planning*, in: Bruton M. ed., *The Spirit and Purpose of Planning*, Hutchinson, London.

Rydin Y. 1998, *Urban and environmental planning in the UK*, MacMillan Press Ltd: London.

Saler W., Faludi A. 2000, *The revival of strategic spatial planning,* Koninklijke Nedelandse Akademie van Wetenschappen [Royal Netherlands Academy of Sciences], Amsterdam.

Serra J. 2001, *Territorial Foresight: More than Planning less than,* EC Conference on Regional Foresight. Dec 13th, 2001. [Online], [Accessed on 21st July 2002]. Available from World Wide Web: http://foren.jrc.es/Docs/Conference/con-prog.htm.

Sharp T. 1940, *Town Planning.* Harmondsworth/Penguin Books, England/New York.

Shipley R. 1998, "Visioning: did anybody see where it came from?", *Journal of Planning Literature,* Vol. 12, No. 4, pp. 407-416.

Shipley R. 2002, "Visioning in planning: is the practice based on sound theory?", *Environment and Planning,* Vol. 34, pp. 7-22.

Steiss A.W. 1974, "Models for the Analysis and Planning of Urban Systems", Lexington Books, London.

Taylor N. 1998, *Urban Planning Theory since 1945,* SAGE Publications, London.

Tosics I. 2003, "A new tool for consultants to influence policy-making? Strategic planning in European cities", paper prepared for the Eura – Eurocities – MRI Conferece "European urban development, research and policy. The future of European cohesion policy". Budapest, 28-30 August 2003.

Wachowicz M. 1999, *Object-Oriented Design for Temporal GIS,* Taylor & Francis, London.

Wachs M. 2001, "Forecasting versus envisioning: a new window on the future", *APA Journal,* Vol. 67, No. 4, pp. 367-372.

Webber M.M. 1963, *The prospects for policies planning,* in Duhl L.J. eds., *The urban condition: people and policy in the metropolis.* Simon and Schuster, New York.

Wheelwright S.C. & Makridakis S. 1985, *Forecasting methods for management, 4th ed.* John Wiley and Sons, New York.

8 The end is nigh ... but are we there yet? Futures and the environment

Graham H May

1 Introduction

For the survival of the human race there is no more important issue than the environment that supports life on the earth. For most of human history it has been tacitly assumed that we can use the environment as an inexhaustible resource and a bottomless pit in which to dump our waste, but at least since Thomas Malthus questioned the ability of the earth to provide sufficient food for a growing population there has been concern that in the way we treat our environment we may be threatening the future of our own existence.

This question has been given increasing attention in the last half-century as a growing number of studies have examined our relationship to the environment and its future. The scope of the issues involved was made clear by the publication in 1972 of the Club of Rome study *The Limits to Growth* (Meadows et al 1972). Using the emerging capacity of computer modelling the study examined the interrelationship of, "five major trends of global concern – accelerating industrialization, rapid population growth, widespread malnutrition, depletion of non-renewable resources, and a deteriorating environment," (p. 21) and concluded that, "if present growth trends... continue ...the limits to growth on this planet will be reached sometime within the next one hundred years"(p. 23)

Since then a series of books and reports have repeated the warning that business as usual will lead to disaster, but despite the growing scientific consensus around the threat from Global Warming in particular, there remain a few dissenters and debate about how to deal with the complex issues involved continues.

2 Implicit futures

Although there are examples of the collapse of civilisations resulting from their inability to live in their environment before the Industrial Revolution; (Diamond 2006) some concerns about the shortage of resources as early as the 17th century; (Peterson del Mar 2006, p. 9) and air pollution in ancient Rome; (Lomborg 2001, p. 163) it is the changes resulting from industrial society that really began to focus attention on the environment. Peterson del Mar suggests that earlier, during the Middle Ages and the early Renaissance, even scientists such as Galileo, Bacon, Descartes and Newton regarded nature as, "a collection of inert materials and mechanistic processes that humans could and should manipulate to further their own ends" (p. 7) Much of the early concern that did emerge in the late 18th and 19th centuries centred on the loss of natural phenomena and the wish to conserve them. The shortage of wood became a concern in several European countries and a developing reverence for nature inspired many leading artists and intellectuals. In drawing attention to these concerns there were few explicit references to the future, but implicitly it was seen as less attractive than the past because valued features of nature were, or would be, no longer in existence. Peterson del Mar (2006, p. 63) for example quotes the American novelist John Steinbeck as complaining that it soon, "would be possible to drive from New York to California without seeing a single thing."

The pressure to conserve, or even preserve, features of the past, both natural and man-made for the future remains an important force within the environmental movement. It has resulted in the establishment in many countries of National Parks and other protected areas and internationally of World Heritage Sites. In the United Kingdom, for example, although other more campaigning environmental groups such as Greenpeace and Friends of the Earth play an important role in influencing policy and hence the future of the environment, organisations such as the National Trust, devoted to the conservation of natural and historic landscapes, have a far larger membership. Peterson del Mar (2006, p. 34) however, suggests that even in the 19th century, "Conservationists pointed out how unregulated logging brought erosion, climate change and poor water," clearly anticipating concerns that would later become central to environmentalism.

An important step in the transition from the environmental future as implicit in the desire for conservation to an explicit view of the future came with the publication of Rachel Carson's book *Silent Spring.* Published originally in 1962 the first chapter of *Silent Spring,* A Fable for Tomorrow, provides a brief description of a mythical town where a catalogue of disasters, each of which had already happened somewhere, come together at an un-stated future time and place. Dire though this predicted future is, like many later studies, Carson suggests that it is not inevitable if a different course of action is followed. She outlines such a course in her final chapter, The Other Road,

3 The explicit future

In presenting a picture of the future given certain assumptions and then offering a possible solution to the problems that would be encountered should that future occur Carson was following in the path of a number of earlier writers including Thomas Malthus (1798). In *An Essay on the Principles of Population*, an issue that remains central to the environmental debate, Malthus argued that food production could only increase at an arithmetical rate but that population increased geometrically. Projecting these rates, and assuming that the then current population of the United Kingdom of 7 million was adequately fed, he showed that although double that number, 14 million, could be fed in 25 years time because food supply would also double, the increase in food supply over the next 25 years would only be sufficient to feed half of the extra population as the growth of population and food supply diverged. This future was not inevitable but his solution was fairly drastic, misery for the poor to restrain their tendency to reproduce, as the only way, he thought, to prevent misery for all. Half a century later Stanley Jevons (1866) in the *Coal Question* considered another issue that also remains important to environmentalists today, the future of resource availability. Assuming that the then present rate of growth in demand for coal in the United Kingdom would continue he concluded that, "we cannot long maintain our present rate of increase of consumption... that the check to our progress must become perceptible within a century of the present time...(and) that our present happy progressive condition is a thing of limited duration." It is remarkable that the conclusions expressed by Malthus and Jevons, that the current trajectory of society cannot continue, are repeated in more recent studies 150 and 200 years later.

4 The collapse of industrial society unless...

Whether it was the influence of *Silent Spring*, as Paehlke (1989) suggests, or the iconic photographs of the Earth from space provided by the Apollo missions that brought the fragility of the planet into focus, environmentalism came to the fore in the 1970s. Although there had been a number of writers throughout the 20th century who questioned the impact that humanity was having on the planet it was only in the late 60s that concern for the future of the environment emerged with any force. Numerous books and official enquiries published since have expressed the opinion that the continuation of the western model of society based on continued economic growth, or business as usual, is unsustainable. Probably the most influential of the early studies was the Club of Rome funded *The Limits to Growth*. Based on the idea that the five areas of concern, population, food production, industrialization, pollution and the consumption of non-renewable natural resources were all increasing exponentially, as Malthus believed of population and Jevons had assumed for coal consumption over 100 years before, the authors concluded that collapse could only be avoided by adopting a series of different policies in each of these areas. But *Limits* was far from the only voice singing a similar song.

In 1969 the United States National Academy of Sciences published *Resources and Man* (National Academy of Sciences 1969) a study focussed on the future supply of material resources as they affected North America. Detailed chapters reviewed the prospects for food and mineral resources from both land and sea and for energy. In view of more recent concerns about "Peak Oil" it is interesting to note that the author of the chapter on energy resources was M. King Hubbert. Accepting the uncertainty of recoverable reserves Hubbert provided two estimates for the peak of world oil production, 1990 and 2000! Overall the study concluded, "It is not certain whether, in the next century or two, further industrial development will be foreclosed by limitations of supply. The biggest unknowns are population and rates of consumption. It is self-evident, however, that the exponential increases in demand that have long-prevailed cannot be satisfied indefinitely. If population and demand level off at some reasonable plateau, and if resources are used wisely, industrial society can endure for centuries or perhaps millennia" (p. 6) In order to achieve the necessary transition the study made 26 recommendations under four headings: Early Action, Policy, Research and Organisation.

A more drastic view of the human predicament was taken in *A Blueprint for Survival*, published by the Ecologist magazine in 1972. "If current trends are allowed to persist, the breakdown of society and the irreversible disruption of the life-support systems of this planet, possibly by the end of the century, certainly within the lifetimes of our children, are inevitable" (p. 1) Food shortages within the next thirty years (by 2002) were considered likely and "present reserves of all but a few metals will be exhausted within 50 years, if consumption rates continue to grow as they are" (p. 4) Silver, gold, mercury, lead, platinum, tin and zinc would all be exhausted before 2000 even if rates of consumption did not increase above then current levels and oil reserves would also have fallen below demand. The decline in resource availability would have major economic consequences including mass unemployment. In order to avoid this collapse a Strategy for Change, Towards the Stable Society was proposed. Beginning in 1975 a wide range of measures aimed at minimising disruption to ecological processes, converting the economy to one of stock rather than flow, stabilising population and creating a new social system would be needed to create the desired network of self-sufficient, self-regulating communities by 2075.

Another plan to save Spaceship Earth was put forward by Paul Ehrlich, whose major concern was overpopulation, and Richard Harriman (1971). They saw the solution in a series of massive changes that had to be made during the next decade. The view that dire though the prospects are, salvation is possible, if we adopt a different path runs through most environmental criticisms of industrial society, even Gordon Rattray Taylor's *The Doomsday Book* (1972) and Gordon and Suzuki's, *It's a Matter of Survival,* (1991) despite the fact that the cover of the former states, "If you think you're going to survive the next thirty years... think again!" and that of the latter, "We have just one decade in which to avert the environmental destruction of our planet." One of the most pessimistic writers, though he did not completely rule out avoiding disaster, was Ronald Higgins, who in *The Seventh Enemy* (1980) outlined a probable future of, "downward drift to an anarchic calamity" by the millennium. After reviewing six threats of population, food, resources, environmental degradation, nuclear abuse and unleashed science and technology, which he suggested could be overcome, he described the political inertia and individual blindness of the seventh enemy, us, as the reason those problems will probably not be solved.

Reason for hope was, however, seen by the group responsible for *The Global Report to the President of the United States* (Barney 1980). Although they concluded that, "If present trends continue, the world in 2000 will be more crowded, more polluted, less stable ecologically, and more vulnerable to disruption than the world we live in now," (p 1) vigorous new initiatives could alleviate these problems. As well as developing projections of population and resources the report considered the likely environmental consequences via impacts on agriculture, water resources, forests, the atmosphere and climate, and species extinction. Among the atmospheric effects acid rain, carbon dioxide emissions and ozone depletion received particular attention.

Despite the fact that we have managed to muddle through to avoid the worst of these predictions of doom by 2000 the doubts about the continued existence of humanity have not died as a number of more recent book titles show. *Beyond the Limits* (Meadows et al 1992), *Earth in the Balance* (Gore 1992) *The Last Generation* (Morgan 1999), *Our Final Century* (Rees 2003), *The 2030 Spike: Countdown to Global Catastrophe* (Mason 2003) and *The Meaning of the 21st Century: A Vital Blueprint for Ensuring our Future* (Martin 2006) for example, all suggest that the environmental and other threats have not gone away and that although we may have made it this far we cannot assume our luck will hold unless we take immediate, collective and decisive action to alter our trajectory.

Most of these authors, although contending that business-as-usual would lead to catastrophe, offered ways out in the form of alternative preferable futures for humanity. Such visions of a better future have a long history in utopian literature, but the increasing concern about the future of the environment, particularly the threat of global warming, has lead to an increasing number of studies aimed at ways of achieving more desirable futures. For example, following the aim of the United Nations Framework Convention on Climate Change to stabilize greenhouse gas concentrations the UK Government adopted the target of reducing CO_2 emissions by 60% by 2050. In *Decarbonising the UK: Energy for a Climate Conscious Future* the Tyndall Centre for Climate Change Research produced five scenarios of energy supply and demand indicating how this might be achieved and the measures needed to bring it about under varying circumstances. (Andrews et al 2005) The International Energy Agency has also produced a Sustainable Development Vision Scenario (OECD/IEA 2003) again outlining how such a desirable future could be achieved.

5 Sustainability

Although the concept of a sustainable future was the basis of the *Blueprint for Survival* in 1972 and Lester Brown's 1981 *Building a Sustainable Society*, the idea did not gain wide recognition until the publication of the Bruntland Report *Our Common Future* in 1987. Brown's approach to sustainability emphasized, "the maintenance and use of the earth's renewable resource base. He stressed the protection of land and soil quality, sustaining biological resources against the pressures of overpopulation and industrialization, the use of renewable energy sources, and the need for population stabilization." (Paehlke 1989, p. 140). The Bruntland Report, following an analysis not dissimilar from that of earlier studies, concluded that, “many development trends leave increasing numbers of people poor and vulnerable, while at the same time degrading the environment” (p. 4) In order to enable both greater global equality and a sustainable environment a new development path was required. That sustainable development path was defined as, “development which meets the needs of the present without compromising the ability of future generations to meet their own needs” (p. 43) This definition, which has become probably the most quoted, is Dresner (2002) contends both simple and vague. This is both its strength and its weakness. The report goes on to make clear that it sees sustainable development as, “ a process of change in which exploitation of resources, the direction of investments, the orientation of technological devel opment, and institutional change are made consistent with future as well as present needs” (p. 9) but as subsequent debates have shown that leaves much open to interpretation.

One of the particular features of the Bruntland Report was its emphasis on the need for equity not just between generations as noted in its definition of sustainable development but also within generations, poverty being seen as an important cause of environmental degradation. Dresner (2002) argues that combining sustainability, which was essentially the concern of the developed world with the ability of the physical environment to support life, with development, the concern of the developing world for greater global equality, was the only way to bring the two groups represented on the Commission together. In doing so he suggests Bruntland was able to overcome the opposition of developing countries who saw the environmental concerns of the developed world as a way of preventing them sharing the benefits of development. It also re-emphasises the inter-relationship between the environment and the economic and political decisions we make. The need to link the environment and the economy in the pursuit of sustainability was the focus of Herman Daly's Steady-state Economy (1977) that like *Blueprint for Survival* emphasised an economy of constant stocks that would minimise the throughput of resources and see improvement in forms other than economic growth.

As Williams (1994) notes, however, Bruntland's approach to sustainability addresses only the human-environment relationship. “Deep ecologists demand that nature is no longer seen solely as raw material for people's benefit. Instead, flora and fauna are viewed as intrinsically valuable” (Williams 1994, p. 172). To an extent this echoes the attitudes of some of the earliest environmentalists and reflects the approach of Buddhist economics in which, “Consumption is less important than creative activity, and conspicuous consumption is openly offensive” (Paehlke 1989, p. 173). It is also the essence of Lovelock's Gaia

Hypothesis, "that views planet Earth...as part of one great organism" (Burrows et al. 1991, p. 218). Although Lovelock, "considers that Gaia will probably survive her present environmental crisis....The survival of the human species is not necessary to the survival of Gaia as a living organism" (Burrows et al., p. 219)

6 Timescale

The Bruntland definition of sustainable development refers only to future generations whose interests are not represented in current decisions and whose options present profligacy are rapidly closing. Such interests could extend indefinitely into the future and as Tonn (1988) contends are difficult to assess, because future generations:
- Are not able to express their preferences and our assessment of them may well be incorrect;
- We do not know who will be alive to constitute the future generations; and
- The difficulties of predicting the future costs and benefits of potential policies are such as to be almost impossible.

Pearce et al. (1989) offered one way to avoid the need to consider the infinite future in suggesting intergenerational equity through which each generation is fair to the next by leaving them an inheritance of wealth no less than they inherited. Even this approach requires a longer view into the future than is found in most disciplines and it is a noticeable feature of most environmental studies that their concern extends to what is generally regarded as the longer term.

Malthus predicted that population would exceed food supply in little over 25 years a relatively short timescale for environmental studies. It is, however, not dissimilar to *The Blueprint for Survival* (1972) and *the Global Report to the President* (1980) which both saw a worsening future by 2000, 28 and 20 years ahead respectively. The millennium, not surprisingly, encouraged many studies of the future from the 1960s on, the earliest looking nearly 40 years ahead, though many were not focussed on the environment. (For a list of some of the 2000 studies published between 1963 and 1991 see May 1999) *The Blueprint,* on the other hand, proposed a programme of actions that extended over a period of 100 years to reach the goal of a sustainable society. Looking a century ahead is not unusual in studies of the relationship between humanity and the environment, *The Limits to Growth,* for example predicting the limits would be reached within the next 100 years from 1972. More recently the forecasts of global warming by the International Panel on Climate Change have also looked as far ahead as the end of the 21st century (IPCC 2000). In a review of 21 studies Greeuw et al (2000 p 88) show that most of them look at least 20 years ahead and several 100. They also concluded that the more recent studies considered longer time horizons than those produced in the early 1990s.

7 Methods

Some early commentators on the relationship between humanity and the environment, such as Malthus and Jevons, relied on the simple extrapolation of existing trends to identify problems ahead. More recent writers have also used extrapolation to emphasise their concerns, for example, The Ecologist (1972) that based its need for change on the projection of exponential growth and Commoner (1972) who suggested that many rivers and lakes in the USA may become incapable of supporting the ecological cycle and that pollution of surface water may create an intolerable threat to human health within the next thirty years. More recently two main methods have dominated, causal modelling and scenarios, which have been used either alone or in combination to explore possible futures.

One of the first uses of models in an environmental context was in *The Limits to Growth* (Meadows et al. 1972). *Limits* used the World 3 model developed from the pioneering work in Systems Dynamics by Jay Forrester at MIT. The authors believed that the model was the only one in existence at the time that attempted to dynamically interrelate variables such as population, food production and pollution at a global scale over a time horizon of more than thirty years. Although aware of its limitations they argued that as a formal mathematical model it had two important advantages, first, that every assumption is precise and open to inspection and second, that using the computer allows the implications to be traced without error. An indication of the complexity of the model is given in the follow up study *Beyond the Limits* (Meadows et al 1992) that refers to the 637 page technical report *Dynamics of Growth in a Finite World.* (p. 240) In both *Limits* and *Beyond the Limits* the model is used to explore alternative futures based on a range of different assumptions.

A series of studies of eleven elements, several incorporating models, prepared by individual national government departments were used to provide projections for the *Global Report to the President* (Barney 1980) there being no suitable integrated model available. Being prepared by different people, at different times, using different methodologies and for different needs made it difficult to integrate them into a coherent picture of the future. It also led to the authors to conclude that, "Analysis of the assumptions underlying the projections and comparison with other global projections suggest that most of the Study's quantitative results understate the severity of potential problems the world will face as it prepares to enter the twenty-first century" (p. 7)

The extent to which models have become important in the study of environmental futures is emphasised by Greeuw et al (2000). They reviewed 13 models, most published in the 1990s, which they divided into three categories:

- "Models that have been built with the explicit objective of providing integrated insight into a broad range of environmental, economic and socio-cultural aspects of sustainability," including World 3
- "Models that put an emphasis on the link between the energy sector and the environment," including the IPCC models and
- Models that focus, "on the link between the economy and the environment."

None of the models reviewed provided the, "full horizontal and vertical integration, policy relevance and European focus," that the authors were seeking but they did provide valuable insights into the problems and possibilities of developing an integrated European model.

Among the groups of models reviewed by Greeuw et al is the International Panel on Climate Change's multi-model of Greenhouse Gas emissions. Six models each with a number of sub models are included in the IPCC approach that attempts to integrate a wide range of factors that affect climate. The factors that are incorporated in the models including, GDP, population, lifestyle developments, energy demand, food consumption, technological change, agriculture, deforestation, land use, atmospheric composition, zonal atmospheric climate, oceanic climate, biosphere and chemistry make their complexity abundantly apparent. As Schneider (2002) notes, "The problem for climate scientists is separating out quantitatively cause and effect linkages from among the many factors that interact with the climate system. It is a controversial effort because there are so many interacting complex sets of processes operating at the same time that debates about the adequacy of the models often erupt" (p 130). There have been rapid advances in modelling since 1990 as greater computing power has become available. In particular Schneider notes the development of transient models that allow the examination of the build up of greenhouse gases over time as an advance on the earlier equilibrium models that could only calculate, "how earth's climate would eventually be altered after CO_2 was artificially doubled and held fixed indefinitely rather than increased incrementally over time as it is in reality" (p. 137).

The IPCC models are used in the development of scenarios for emissions and their implications for changes in global climate and the rise in sea level. In an environmental context, as elsewhere, scenarios are used in an attempt to encompass the uncertainty inherent in forecasting the future, particularly so far ahead as many environmental studies attempt to look. The IPCC first developed a set of long-term emission scenarios in 1990 and 1992. They were extensively used in the, "analysis of possible climate change, its impacts, and options to mitigate climate change," but, "changes in the understanding of driving forces of emissions and methodologies," led to the decision in 1996 to develop a new set. (IPCC 2000, p. 3). Four different storylines based on varying demographic, social, economic, technological and environmental developments were created to describe the relationship between emission driving forces and their evolution. Six models were then used to develop 40 scenarios that encompass the current range of uncertainties relating to future emissions and the driving forces. The 40 scenarios are divided into six scenario groups drawn from the original storylines and although all scenarios are considered equally valid six illustrative scenarios, one for each group, have been used in most of the subsequent analysis. (For details see IPCC 2000)

In addition to the IPCC scenarios Greeuw et al. (2000) reviewed twenty-one other scenario studies prepared or under development. The studies were classified into two design approaches: analytical desk studies, themselves either quantitative (model based) or qualitative, and participatory. The scenarios within the studies were further analysed according to the dominant driving forces; Social, Environmental or Economical; and their overriding character; Wait and See (limited policy action), Just do it (considerable policy intervention), Doom Monger (a pessimistic outlook), and Carpe diem (a positive outlook). The review offered a number of lessons including the value of participation in scenario development, the integration of surprise and different methods, the inclusion of different perspectives and the need for the balanced integration of environmental, social, economic and institutional processes. It also concluded that, "If a scenario study starts with a model, the narrative resulting from this modelling exercise almost always becomes too narrow. However, when a study starts with the narrative the model cannot include all the richness of the narrative" (p. 91). Even with these more sophisticated methods exploration of the environmental future is far from perfect.

8 The sceptics

Although most views of the future from an environmental perspective express concern about the relationship between humanity and the environment, and in particular the ability of the environment to support life if we continue on a business-as-usual trajectory, other voices, mostly from different disciplines, have challenged this approach. The contrasting opinions were explored in a debate that took place in New York in October 1992 between Norman Myers and Julian Simon subsequently published as *Scarcity or Abundance: A Debate on the Environment* (Myers and Simon 1994). Simon had earlier published a response to *The Global Report to the President* with Herman Kahn (Simon and Kahn 1984) that contended that, "If present trends continue the world in 2000 will be less crowded (though more populated) less polluted, more stable ecologically and less vulnerable to resource-supply disruption than the world we live in now" (p. 1) Although they did not predict, "that all will be rosy in the future," (p. 3) they argued that the trends they examined were, "improving rather than deteriorating". "We are confident that the nature of the physical world permits continued improvement in humankind's economic lot in the long run, indefinitely," because men and women, "will address problems with muscle and mind, and will *probably* overcome, as has been usual throughout history" (p. 3). They were, however, less optimistic about the constraints imposed, "by political and institutional forces, in conjunction with popularly-held beliefs and attitudes about natural resources and the environment" (p. 3).

Simon had also offered a bet in 1980 that any raw material, to be picked by his opponents, would drop in price in real terms. The bet was taken up by a group led by Paul Ehrlich. They selected a number of resources: chromium, copper, nickel, tin and tungsten, over a period of ten years. By September 1990, "not only had the total basket of raw materials but also each individual raw material had dropped in price" (Lomborg 2001, p. 137). Myers (Myers & Simon 1994, p. 99) contended that the fall in prices could be explained by the unusual circumstances of the downturn in industrial growth during the 1980s and by other distortions in pricing, classic short-term market variations that do not reflect longer-term resource issues. Simon refuted the claim and offered a similar bet that Myers did not take up. (Myers & Simon 1994, p. 206). The problem Lomborg (2001) suggests is that, "Should you be inclined, you could easily write a book full of awful examples and conclude that the world is in a terrible state. Or you could write a book full of sunshine stories of how the environment is doing ever so well. Both approaches could be using examples that were absolutely true" (p. 7). Lomborg in fact set out to challenge Simon's view (p. xix) but concluded by examining global data that on average, despite local variations, the overall situation was improving, in contrast to the opinions of most environmentalists with which he had previously agreed.

9 Some concluding thoughts

The future of the environment has been a concern for over 50 years, at least since Rachel Carson published *Silent Spring*. In that time a wide range of issues have been raised including many forms of air pollution, such as acid rain, water pollution and supply, waste, the impact of chemicals, threats to biodiversity, food supply, and more recently global warming and climate change, each being of particular concern at certain times. The complexity involved in thinking about the environmental future is further emphasised by the links that each of these issues have to economic, technological, social and political developments. Many of these issues have been brought to the fore by interest groups campaigning for particular policies to be adopted and although as Paehkle (1989) notes environmentalism, "is not an ideology of self-interest," (p. 7), it is, "largely political and ideological in its perspective" (p. 21), questioning, "the logic of private investment decisions and the conventional treadmill model of production expansion to generate economic growth" (p. 22).
As such the projections produced by environmentalists have often been accused of scare-mongering, being overly pessimistic in order to emphasise their concern and convince decision-makers to adopt the sometimes draconian measures proposed. A perhaps surprising feature of many of these forecasts, exemplified by *Limits to Growth* and more recently the projections for global warming, is that they are made to be wrong. It is usually assumed that forecasts are made to provide an accurate picture of the future, but probably the last thing that the authors of *Limits* or the forecasters of increasing CO_2 emissions want is for their forecasts to come true! Their intention is to encourage change in human behaviour that will prevent those dire predictions from happening. With few exceptions environmentalists appear to believe that human induced catastrophe is not inevitable and that we can avoid that future if we only take the necessary precautionary steps. There is some evidence for this and to support Simon's view that we can solve problems and avoid predicted disasters, for example the banning of CFCs in the light of concerns about ozone depletion. On the other hand it is salutary to note that predictions based on the measurement of CO_2 in the atmosphere at Mauna Loa in the 1960s that forecast an exponential increase to 380 parts per million by 2000 were only six years out, that level being reached in 2006 (Rohde 2006) and despite Kyoto and much concern that this is a major cause of global warming there is little sign that the increase is about to cease.

References

Andrews A., Shackley S., Mander S. & Bows A. 2005, *Decarbonising the UK: Energy for a Climate Conscious Future,* Tyndall Centre for Climate Change Research.

Barney G.O. 1980, *The Global Report to the President of the United States: Entering the 21st Century, Volume 1 The Summary Report, Special edition with the Environmental Projections and the Government's Global Model,* Pergamon. New York.

Brown L. 1981, *Building a Sustainable Society,* W W Norton, New York.

Burrows B., Mayne A. & Newbury. P 1991, *Into the 21st Century: A handbook for a Sustainable Future,* Adamantine,Twickenham.

Carson R. 1965, *Silent Spring,* Penguin Books, Harmondsworth.

Commoner B. 1972, *Impending Environmental Catastrophe* in Fowles J. ed., *Handbook of Futures Research* p 605-616, Greenwood, PressWestport.

Daly H. 1977, *Steady-State Economics,* Freeman, San Francisco.

Diamond J. 2006, *Collapse: How Societies choose to fail or survive,* Penguin Books, London.

Dresner S. 2002, *The Principles of Sustainability,* Earthscan Publications, London.

The Ecologist 1972, *A Blueprint for Survival,* Vol 2, No 1, January.

Ehrlich P.R. & Harriman R.L. 1971, *How to be a Survivor: A Plan to save Spaceship Earth,* Ballantine Books, London.

Gordon A. & Suzuki D. 1991, *It's a Matter of Survival,* Harper Collins, London.

Gore A. 1992, *Earth in the Balance: Forging a New Common Purpose,* Earthscan Publications, London.

Greeuw S.C.H., van Asselt M.B.A., Grosskurth J., Storms C.A.M.H., Rijkens-Klomp N., Rothman D.S., and Rotmans J. 2000, *Cloudy Crystal Balls: An Assessment of recent European and global scenario studies and models,* Copenhagen, European Environment Agency, available at http://reports.eea.eu.int/Environmental_issues_series_17/en/envissue17.pdf accessed 11 January 2004.

Higgins R. 1980, *The Seventh Enemy: the Human factor in the Global Crisis,* Pan Books, London.

International Panel on Climate Change 2000, *IPCC Special Report: Emissions Scenarios: Summary for Policymakers* IPCC available at http://www.ipcc.ch/pub/sres-e.pdf accessed 24 January 2007

Jevons S. 1866, *The Coal Question: An Inquiry Concerning the Progress of the Nation and the Probable Exhaustion of Our Coal Mines,* Second Edition, London,

Macmillan, available at http://www.econlit.org/lbrary/YPDBooks/Jevons/jvnCQ.html accessed on 11 January 2007.

Lomborg B. 2001, *The Skeptical Environmentalist: Measuring the Real State of the World*, Cambridge University Press, Cambridge.

Malthus T. 1798, *An Essay on the Principles of Population as it reflects the future improvement of society with remarks on the speculations of Mr Godwin, M. Condorcet and other writers*, London, Jackson, available at http://www.gutenberg.org/dirs/etext03/prppl10.txt, accessed on 11 January 2007.

Martin J. 2006, *The Meaning of the 21st Century: A Vital Blueprint for Ensuring our Future*, Eden Project Books, London.

May G.H. 1999, *After the party is over: Futures Studies and the Millennium* in Didsbury H F Jr (ed) *Frontiers of the 21st Century: Prelude to the new millennium*, World Future Society, Bethesda.

Meadows D.H., Meadows D.L., Randers J. and Behrens W. W. III 1972 *The Limits to Growth*, Earth Island, London.

Meadows D.H., Meadows D.L., & Randers J. 1992, *Beyond the Limits: Global Collapse or a Sustainable Future*, Earthscan Publications, London.

Myers N. and Simon J.L. 1994, *Scarcity or Abundance: A Debate on the Environment*, W W Norton, New York.

National Academy of Sciences 1969, *Resources and Man: A Study and Recommendations by the Committee on Resources and Man*, W H Freeman, San Francisco.

OECD/IEA 2003, *Energy to 2050: Scenarios for a Sustainable Future*, International Energy Agency, Paris.

Paehkle R.C. 1989, *Environmentalism and the Future of Progressive Politics*, Yale University Press, New Haven.

Pearce D., Markandya A. & Barbier E.B. 1989, *Blueprint for a Green Economy*, Earthscan Publications, London

Peterson del Mar D. 2006, *Environmentalism*, Pearson Books, Harlow

Rohde R.A. 2006, *Atmospheric Carbon Dioxide Measured at Mauna Loa, Hawaii* available at http://en.wikipedia.org/wiki/Image:Mauna_loa_Carbon_Dioxide.png accessed 11 January 2007.

Schneider S.H. 2002, *Modeling Climate Change Impacts and Their Related Uncertainties*, Chapter 5 in Cooper R.N. and Layard R. *What the Future Holds: Insights form Social Science*, The MIT Press, Cambridge.

Simon J.L. & Kahn H. eds. 1984, *The Resourceful Earth: A response to Global 2000*, Basil Blackwell, Oxford.

Taylor G.R. 1972, *The Doomsday Book*, Panther Books, London.

Tonn B.E. 1988, Philosophical Aspects of 500 year Planning, *Environment and Planning A*, Vol. 20, pp.1507-22.

Williams C.C. 1994, *From Red to Green: Towards a new antithesis to capitalism?* Chapter 11 in Williams C.C. & Haughton G., *Perspectives Towards Sustainable Environmental Development*, Avebury, Aldershot.

The World Commission on Environment and Development 1987, *Our Common Future*, (The Bruntland Report), Oxford University Press, Oxford.

9 Managing the future

Patrick van der Duin and Erik den Hartigh

1 Introduction

Scholars in the management sciences study the behavior of companies. In this chapter we explore how the future is being addressed in the innovation and strategy domains of the management sciences.[1] While not as explicitly as futures research itself, the management sciences, and these two domains in particular, are strongly related to the future. For the management sciences, outcomes of futures research are not goals in themselves but provide inputs to decision-making processes within companies. To make sure that futures research is appropriately used in these decision-making processes, researchers and managers should take into account the type of decisions for which it is used. In this chapter, we focus on decisions regarding strategy and innovation.

Innovation is strongly related to the future. Innovation processes are about creating new products, processes or technologies. These processes provide ample opportunities for incorporating new (future) developments. A promising idea for an innovation, based on an envisioned future, can be countered or supported by future changes in technology, economy, or society. On the one hand, certain future expectations might turn out not to be true and replaced by others. On the other hand, unforeseen future developments can enhance the probability of an innovative idea to be realized. The implication of all this is that every innovation process should take the future into account.

The numerous wrong predictions of market diffusion of new products and services illustrate that the use of futures research in innovation processes is not without problems. A *technology push* approach, for example, ignores the

1 We do not regard innovation as a type of strategy but as an implementation of a specific strategy.

influence of important societal and market developments on the acceptance of innovations. In the other extreme, the *market pull* approach, only taking into account societal and market developments limits the technological window of opportunity for innovation and runs the risk of missing out on new technological possibilities. It appears that for companies it is essential to carefully balance the nature of the innovation process and the way futures research is carried out and used in this innovation process.

Strategy focuses on how organizations should behave, organize, adapt or learn to survive. Every strategy by definition, implicitly or explicitly, defines and deploys a future vision towards the environment and the perceived role of the company within this environment. Therefore, every strategy process can benefit from futures research. The way futures research is carried out should be in line with the strategy process. The diversity in types of strategy processes (e.g., outside-in, inside-out, static or dynamic) should be mirrored by the specific futures research process. For example, with outside-in approaches the contribution of futures research is dominantly to identify environmental trends, while with inside-out approaches the focus is on visioning how the company's strategic intent can be realized. With static approaches, the contribution of futures research is often limited to singular predictions, while with dynamic approaches futures research delivers a more continuous contribution.

2 Strategy

In this section we provide an illustration of how different strategic schools of thought deal with the future. In the field of strategic management, researchers and managers can take different basic perspectives on their subject matter. These basic perspectives are known as *theories of the firm*, because they provide explanations about the nature of the firm, i.e., why it exists, about its behavior, about its relation to the market and about its performance. Multiple *theories of the firm* exist in parallel and in this chapter we will consider only a few of the most important *theories of the firm*. We distinguish them by two dimensions:

1) Does the reasoning start from the market or from the company?, i.e., is the future influencing the company or is the company influencing the future?,
2) Is the reasoning basically static or dynamic?, i.e., is the future linearly predictable, or is the future continually created?

In doing this, we use the perspectives as defined by Teece, Pisano & Shuen (1997) and we classify them as done by Zegveld (2000), see figure 1.

Figure 1 A classification of strategy perspectives (source: Zegveld 2000)

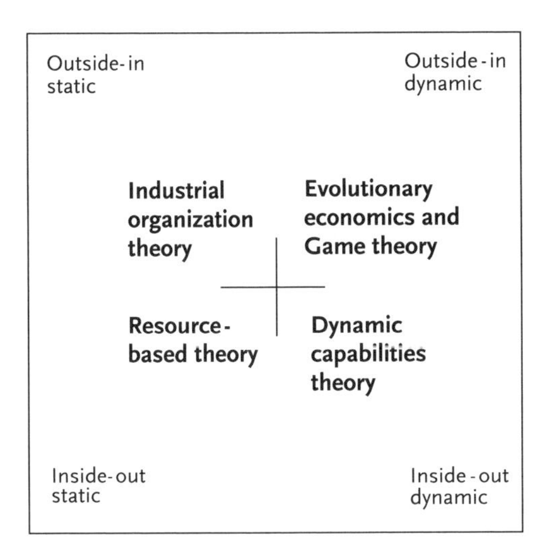

The first dimension for distinguishing theories of the firm is by the order of their reasoning: starting with the market, i.e., outside-in, or starting with the company, i.e., inside-out.

2.1 *Outside-in, static*

Starting with the market implies taking an outside-in approach. The best-known perspective of this approach is the *industrial organization theory of the firm* (Bain 1959; Porter 1980, 1985; Scherer & Ross 1990). The basic reasoning is that the structure of the market determines the behavior of the companies in that market, and that the competition between the companies determines their performance. Companies make a deliberate choice in which market they want to compete and they try to establish a favorable position within this market. Such a position will enable them to realize a cost advantage relative to their competitors, i.e., *cost leadership*, or to differentiate themselves from their competitors, i.e., *differentiation*.
To take strategic decisions using the industrial organization perspective, it is important to have information on the future market developments and on how the company can compete given these market developments. Important tools that can be used to do so are (see also Zegveld 2004): industry analysis, technology analysis, Porter's 5-forces model, SWOT analysis, and value chain analysis.

When the future has been more or less accurately 'predicted' by using these tools, management can decide what actions the company will take to become as profitable as possible, i.e., the industry segment it wants to be in, the strategic positioning, the products and services it will market and the investments it will make.
It is important to recognize that the reasoning of the industrial organization perspective is essentially *static*: the future market developments are predicted, a strategic plan is made, actions are taken and finally the outcomes become visible in the company's performance, i.e., profits, cash flow or products sold. In the meantime, it is assumed that the market does not change in discontinuous ways.

2.2 *Inside-out, static*

Starting with the company implies taking an inside-out approach. The best-known perspective in which this approach is taken is the *resource-based theory of the firm* and its more modern successors like the *core capabilities* or *core competence* perspectives (Penrose 1959; Wernerfelt 1984; Prahalad & Hamel 1990; Barney 1991; 1997). The starting point for reasoning in this theory is the company's strategic intent or ambition. Based on this ambition, the company builds competencies and accumulates resources that are unique and difficult to copy by competitors.

To take strategic decisions using the resource-based perspective, it is important to have information on the company's strategic intent, on its assets and capabilities and how these can be deployed to maximum advantage. Important tools that can be used to do so are (see also Zegveld 2004): strategic intent analysis, technology analysis, competency analysis, asset analysis and strategic architecture analysis.
When the company's strategic intent has been determined and the resources and capabilities have been identified, management can decide which existing resources and capabilities to build upon, which new ones to acquire or invest in, and subsequently which activities to undertake, which customers to target and which markets to serve. Here, too, we should realize that the approach is essentially *static*: once the plans have been made, strategic intent is not supposed to change in major ways and the value of assets and capabilities is not supposed to change abruptly.

In both perspectives discussed above, outside-in and inside-out, it is *assumed* that there is one, relatively predictable future. Of course, no management scientist is naive enough to believe that the future is indeed one-dimensional and predictable, but in these perspectives it is assumed to be at least predictable enough for the company to implement its strategy, test its effects on performance and when necessary change it. It is highly questionable whether in today's turbulent world there is any market or situation where such predictability exists. And yet, many corporate strategic planning processes still build on these static perspectives. And many management scientists still dominantly use those perspectives for their research. An important reason for doing so is that designing and analyzing company strategy becomes progressively more difficult when we take dynamics and uncertainties into account. We know that the use of a static model will at least provide us with some basic information and we are all too happy to put off the dynamic aspects of our corporate strategy to the implementation phase. Part of the problem is that frameworks and methods for dynamic planning or for dynamic research are not yet fully developed.

2.3 Outside-in, dynamic

In the dynamic perspectives, too, we make a distinction between inside-out and outside-in. Evolutionary economics and (evolutionary) game theory are the main perspectives in which an outside-in approach is used and in which dynamics are taken explicitly into account (Hannan & Freeman 1977; 1989; Nelson & Winter 1982; Levinthal 1991; Brandenburger & Nalebuff 1995).

Scholars using evolutionary economics study the process of evolution of business populations and the ways in which individual companies are fit enough to survive the environmental selection forces by adapting to the environment. In this perspective, the future is not something static that companies can easily adapt to, but it is fickle and changing. It requires not *adaptation*, but rather *adaptability*. Companies that are not able to continually adapt will be forced out of business.

Game theory, especially in its evolutionary variant, is a perspective in which scholars analyze how interactions between individual players affect collective outcomes of this interaction. This implies that future outcomes are not seen as uniquely predictable, but as depending on which actions the players in the game will take. For example, in the well-known *prisoner's dilemma*, a single-round game of 2 players that can each take 2 actions, there are 2 x 2 = 4 possible futures. Increasing the number of rounds in the play, i.e., evolutionary game theory, increasing the numbers of players or increasing the numbers of actions each player can take, exponentially increases the number of possible

futures. The corporate strategy interpretation of game theory starts from the *value net*, which provides an identification of the players and the interactions between those players. The next step is identifying the elements of the game and finding opportunities to change the game, e.g., by changing the players, changing the rules of the game or changing the scope of the game.
To take strategic decisions using these perspectives, it is important to have an idea of how interaction between players shapes the future and of how this future in turn influences the possible actions the company can take. Important tools that can be used to do so are (see also Zegveld 2004): population ecology analysis, adaptability analysis, value net analysis, cooperative game analysis and non-cooperative game analysis.
In recent years these tools and the research fields from which they stem became increasingly popular. In organization population ecology and selection, research has been conducted as to how the environmental selection processes work. A particularly promising approach is that of computational analysis and agent-based computer simulation (Epstein & Axtell 1996). These relatively novel methods provide relatively easy ways to explore multiple futures and to explore how a company can adapt to these futures and even influence or create them. The most important challenge in this field is to create a deep managerial understanding to match the analytic rigor of the methods.

From the field of organizational learning, another promising stream of analysis has sprung forward, namely that of scenario planning (De Geus 1997; Van der Heijden 1996). From experience with strategic planning at Royal Dutch/Shell it emerged that conventional strategic planning methods like those advocated by Ansoff (1965) were no longer sufficient. In the 1980's, Arie de Geus (1997) and Kees van der Heijden (1996) at Royal Dutch/Shell's strategic planning department started experimenting with scenario planning, not longer assuming a single future for the company, but rather a portfolio of different futures.

2.4 *Inside-out, dynamic*

In the past twenty years, quite a lot of research was conducted using the resource-based view. While the resource-based view was still in its infancy (e.g., Newbert 2007), a need for a more dynamic version of this view appeared. The article *Competing for the future* by Hamel & Prahalad (1994) already pointed in this direction. In 1997, Teece, Pisano & Shuen (1997) coined the term *dynamic capabilities*, which they define as *"the firm's ability to integrate, build, and reconfigure internal and external competences to address rapidly changing environments"* (p. 516). The dynamic capabilities perspective is an extension of the resource-based view, which, while still departing from the company's capabilities, explicitly takes into account the paths that have created these capabilities and the future paths open to the company (Teece, Piano & Shuen 1997; Eisenhardt & Martin 2000; Helfat & Peteraf 2003). Core notions are dynamics, evolution, change and path-dependence. The company should develop dynamic capabilities, which means is has to be able to renew existing resources or competences. In this way, the company will become adaptable and therefore able to address a continually changing environment. This is difficult because of *path-dependence*: current capabilities and positions are heavily dependent on resource commitments made in the past. In the same way, commitments made today will impact which possibilities remain for the future.

In this perspective, the future is looked upon as being shaped in the interaction between the company and its environment. For the concept of future this means that, on the one hand, there is no single, predictable future, but on the other hand, path-dependency ensures that the range of possible futures is not endless. To take strategic decisions using the dynamic capabilities perspective, it is important to have information on the company's existing resources and competences and on how these can be changed or further built on, on its positions, on how these positions came into being and on how they can be built out, and on the paths that are open for the future. Important tools that can be used to do so are (see also Zegveld 2004): strategic intent analysis, dynamic capability analysis, path-dependency analysis, technology regime analysis and company innovation system analysis.

On some of these tools and methods, such as adaptive capabilities, path-dependence and technology regime analysis, an increasing body of knowledge is becoming available. Such research delivers the basic knowledge needed to understand how on the hand the future can be understood as dynamic and inherently unpredictable, and on the other hand it can be understood as limited in possibilities, because markets and companies can become locked in to a single technological regime or be stuck with their existing capabilities for extended periods.

To summarize the place of 'the future' in strategic thinking figure 2 illustrates which place the future has in different strategic perspectives. A static perspective is in line with a stable and (thereby) predictable future, while outside-in is associated with more change and with much less predictability. Static and dynamic refers to the source of change and thereby uncertainty which can be located within the organization or within its environment.

We have seen that the static perspectives on corporate strategy are becoming less and less relevant in today's dynamic world. This is true for academic analysis as well as for practical strategy making. The dynamic perspectives on corporate strategy have, until now, received considerably less attention than the static perspectives. The body of knowledge as relevant to scholars or managers is indeed growing, but is still relatively small. This causes many scholars and managers to stick to the well-known static perspectives. There is therefore a huge need to adopt perspectives that recognize not just one, predictable future, but a whole range of possible futures. This need for more dynamic approaches of corporate strategy was identified by scholars from Wharton School in 1997 (Day, Reibstein & Gunther, 1997), but it still holds today.

Figure 2 The place of the future in strategic thinking

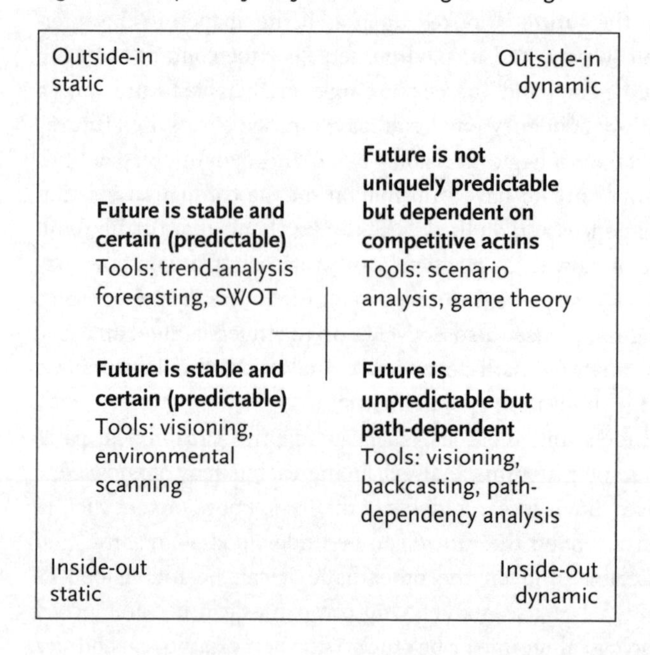

3 Innovation

Innovation has a long history and can be considered one of humanities' oldest activities. From a modern perspective we see that the first R&D-labs were established at the end of the 19th century in Germany. These R&D labs were part of the chemical industry and were soon followed by R&D-labs in the U.S.A. (Bassala 2001).

Innovation was put on the academic agenda by the Austrian economist Joseph Schumpeter who viewed innovation as one of the main causes of economic growth. He introduced the notion of 'creative destruction', the constant renewal of products and services. Originally, Schumpeter thought that this process was mainly executed by small companies (his Mark 1-model) who developed innovations by creating 'Neue Kombinationen', or new combinations of inputs from labor, technology, and capital. Later on he changed his mind and attributed most innovations to large, firms operating in oligopolistic industries (the Mark 2-model). Schumpeter's ideas were not directly taken up by economist who opened the 'black box' only years later. But management scientists discovered more quickly the importance of Schumpeter's ideas on innovation and technology development for the wellbeing of companies.
The current big attention for innovation by management scientists, companies, and governments has its origin just after WW-II when Vannevar Bush, Director of the Office of Scientific Research and Development, presented his influential report entitled *Science, The Endless Frontier* (Bush 1945). In this report Bush states that scientific research is the basis of technological growth, and hence economic growth and an increase in military power. Vannevar Bush' ideas were the start of a series of different innovation management *practices*, each of them dominant in a specific historical period. These *generations* of innovation management are (based on: Liyanage et al. 1999) Miller 2001; Niosi 1999; Rothwell, 1994; Roussel et al. 1991):

Table 1 The four generations of innovation management.

Generations:	Principles of innovation management:
1950 – mid 1960s:	technology push, linear innovation processes
mid 1960s – early 1970s:	market pull, linear innovation processes
early 1970s – mid 1980s:	combined market pull and technology push, coupled innovation processes (with feedback loops)
mid 1980s – early 2000s:	innovation in networks and systems, 'open innovation' (Chesbrough 2003)

This historical development of innovation management can be characterized as *evolutionary* in which a new generation is adjusted to its economic and societal environment thereby attempting to overcome the drawbacks of each former generations, making innovation more complex (Ortt & Van der Duin, submitted). A future perspective on this development by different generations might be motivated by looking for clues for a fifth-generation. And although some authors claim that there is such a fifth (Rothwell 1994), and even a sixth (Yakhlef 2005) generation, these attempts to innovate innovation management do not add sufficiently to speak about a new innovation management *paradigm*. A few recent cases on how large companies such as Shell and Philips manage their innovation process (Verloop 2006; Van Tol, Elst & Smits 2006) suggest that within a single organization different types of innovation management exist. It shows that those companies decide on applying those (existing) innovation management principles on the basis of their organizational needs thereby creating a kind of *contextual innovation management* (Ortt & Van der Duin, submitted). The historical development of innovation management seems to come to a hold and 'new' ways of innovating will mean adjusting old practices to specific organizational contexts.

Innovating means coping with uncertainty, not only the uncertainty of the innovation itself but also uncertainty caused by the fact that the innovation will be implemented and used in a future time, which most of the time is uncertain. One should be aware that innovation processes take time, often a lot of time. Despite the rhetoric around shorter *time-to-market*, the total *time-to-innovation* is still considerable, and often not decreasing. The *time-to-innovation* varies per type of innovation. An incremental innovation, i.e., a slight change or improvement of a product or service, such as a new version of specific software, might take a couple of months to a year. But a more radical innovation, with changes in may aspects of the innovation, can take much longer. For instance, a new car takes seven years to develop and a new medicine somewhere between 10 and 15 years. A consequence of these long innovation development times is that a current, *potential* good idea for an innovation might turn out to be a bad idea at some future date when the innovation is implemented in a market. That is, between the first idea for an innovation idea and its future implementation many changes might take place in society, market, technology, and economy that could counteract the original idea for an innovation that had some high hopes (Van der Duin 2006). On the other hand, some future changes might also support the original innovation idea. But in either case, innovation and future are closely connected, almost in a natural way.

Developing new products and services is breaking with old habits, getting rid of the company's old portfolio. In a way, the future, or at least the company's future, is being shaped by all those innovations. Taken together, the innovations form a kind of *transition path* to an envisioned future. Stated as such, one single future is then being predicted and functions like a kind of *bait*.

The image of the future is meant to inspire innovators. The development of these innovations is then the answer to the question how a future must be realized. The question can also be different: is the innovation we are developing fit for different futures? This question is asked from a more modest stance towards the future. He who asks this question has less faith in shaping the future by himself or his organization but regards the future as given, not by him but by other stakeholders whom he is not able to influence sufficiently. Some studies, indeed, show that being focused on the future is a success factor for being innovative (e.g., Brown & Eisenhardt 1997; Preez & Pistorius 1999; Johannessen, Olaisen & Olsen 1999).
The balance between innovation and futures research is clearly illustrated by the historical development of both topics, as table 2 shows (from Van der Duin 2006; see also Berkhout et al. 2007):

Table 2 Generations of innovation management and futures research

	Innovation processes:	Futures research:
Generation 1: '50 - '60	Technology push	Technology forecasting
Generation 2: '60 - '70	Market pull	Technology assessment
Generation 3: '70 - '80	Coupled innovation processes	Explorative futures research
Generation 4: '80 - present	Innovation in systems or networks	Networked or systemic futures research

Table 2 illustrates that on a general level innovation management and futures research have a parallel historical development. It is difficult to find out which discipline influenced the other, but the resemblance is so striking that it can be assumed that futures research is better used for innovation when both disciplines are mirrored. So, an innovation process should use a technology forecast and in an innovation system futures research should be an 'open' process involving many different inputs from futures researchers from different organizations involved. As a consequence, the concept of *contextual innovation* is also applicable to futures research: *contextual futures research* whereby the decision of how to carry out futures research is determined by to which type of innovation process it used.
Innovation management represents the general principles of addressing innovation at an organizational. On a lower level of detail we have innovation

processes where the specific activities of an innovation *project* are described. Often these innovation processes are pictured as a linear process consisting of different subsequent phases. In each phase innovation futures research (or technological forecasting) plays a different role. Table 3 is from Twiss (1992):

Table 3 The relationship between forecasting and the technological innovation process (Twiss 1992, p. 21).

Phase of the innovation process:	Technology forecasts:		
	Importance	Accuracy	Financial effect of forecasting error
1. Idea generation	High	Medium	Low
2. Technical feasibility	High	Medium	Low
3. Design & development	Low	High	Medium
4. Preparation for production and marketing	Very low	High	High
5. Post launch	Not applicable	Not applicable	Not applicable

This table regards futures research mainly as forecasting. In the first two phases, i.e., idea generation and technical feasibility, forecasting plays its most important role. These phases are the farthest away from market implementation which makes a view in the future valuable. Because in these phases not much (quantitative) information is available the level of accuracy of the prediction is not high. Luckily, in these phases the financial impact of a wrong forecast is not that high either. Given that the level of investments in the innovation increases when its development process comes closer to market implementation the importance of accurateness will increase accordingly. Phases 3, 4, and 5 benefit less from forecasting because most decisions about the innovation are then already taken, making these phases more operational than strategic. To sum up, forecasting in particular and futures research in general will mainly be of importance in the first phases of an (linear) innovation process than in the latter ones where market research methods will be more popular.

Both innovation and future can also be linked to how employees in a company are related to each other. That is, a case-research by Ybema (2004) about a Dutch amusement park called 'De Efteling' reports about a schism between its employees. One group considers itself the founders of the company, proud of establishing it and of bringing it to the successful position it currently holds. The members of the other group have a shorter career at 'De Efteling', they come from different other companies. They are more focused on how to change the company and how to bring it into a new phase of its existence. Ybema calls this difference as a clash between 'nostalgia' and 'postalgia'. A difference between those who stick to the past, those who do want to make big changes and those who are happy with the *status quo*, and those who want to prepare the company for a different future, who consider a day without an innovation as a day gone, and who want to keep up with the big changes they witness around the company. It is difficult to find a solution to the problems that are caused by this schism. After all, a divided company will not be a very efficient one as well as a not innovative one. Maybe it is, nevertheless, important to have both camps within a company since they could form each others check and balances. That is, a company in which the past is the dominant mental map and in which people fear change will, by definition, have no future. On the other hand, a company inhabited by employees who are obsessed with constantly changing almost every aspect of the company driven by every change in future expectations, will not hold a steady course to the future and might be in danger of going way out of line. So, bearing in mind the huge importance of thinking about the future and acting upon it does not mean that we get our future view blemished by every (possible) change in our environment. It takes a steady arm to sail the toughest seas.

4 Concluding remarks

We conclude that the quality of futures research is not only decided by how it is carried out. It is equally important that futures research is connected to the perspective and the type of decision-making processes it is intended to feed. The worst thing a futures researcher might hear is not that his work is not up to standards but that his work is simply not relevant to the company and its decision-making process. It is therefore crucial to relate the type of futures research to different perspectives and types of decision-making processes within organizations, both for the decision-maker and for the futures researcher.

For futures researchers improving their work not only means improving futures research methods and processes, but, maybe even more important, connecting their work to the decision-making processes it is meant to support.

Therefore, futures researchers need to understand not only the *immediate* context of their work, i.e., the decision-making process, but the *organizational* context, i.e., the innovation and strategic issues, of these decision-making processes as well. A better conceptually understanding of the principles of management science can be considered a fruitful way to improve the quality futures research.

In the discussion on the role of futures research in strategic management, we saw that there is a huge need for a further development of the dynamic perspectives, not just as fancy new tools for academic analysis, but also as practical tools to be used by managers. Futures research that is on the one hand analytically rigorous and on the other hand works in direct interaction with management, seems extremely suited to connect to such developments. In this way, futures research may deliver a valuable contribution to dynamic views on strategic management, in which the future is seen as changeable and multi-dimensional rather than static and predictable.

In the discussion on the role of futures research in innovation, we saw that innovation and the future are related to each other in an almost natural way. Hereby the future can function as a source of inspiration to innovators. The close relationship between futures research and innovation is also illustrated by their historical development which shows many similarities. On an innovation process level, futures research adds the most value in its first phases when not much information about the innovation is present and when the innovation is still a long time from its implementation. Attitudes towards the future can also have organizational impact, as the example of 'De Efteling' shows. Being focused on the future and being focused on innovation can, therefore, be considered as two sides of the same coin.

Both strategy and innovation are important aspects of companies' decision-making processes. They are related to the future in an almost natural way, which is to some extent true for the management sciences as a whole as well. Doing business and being successful in doing business requires an above average attention to the future, how it develops or how it can be created. Successful businessmen and -women possess a vision of the future and they know instinctively how to anticipate on future developments or how to create opportunities. Often they will understand and analyze the future in an implicit way, acting by 'gut feel' instead of using the academic *methods* of futures research.

It should be noted that companies will not automatically take a long-term view of the future or that futures researchers do not have to convince managers to systematically analyze the future. Quite the contrary: the increasing influence of shareholders, requiring companies to provide quarterly reports of their earnings, often results in a more short-term view of the future. There seems to be an increasing focus on short-term profits, sometimes even placing a burden on reaching long term goals. *"The urgent drives out the important"*, seems to be some shareholders' favorite Henri Kissinger-quote.

The 'battle for the future' is on and can only be won if futures researchers can clearly demonstrate the added value of their methods for analyzing the future, added value not only as academic tools for abstract analysis, but also as a contribution to generating *sustainable competitive advantage.*

References

Ansoff, H.I. 1965, *Corporate strategy: an analytical approach to business policy for growth and expansion*, New York, McGraw-Hill.

Bain, J.S. 1959, *Industrial organization*, New York, John Wiley & Sons.

Barney, J.B. 1991, "Firm resources and sustained competitive advantage", *Journal of Management*, Vol.17, pp.99-120.

Barney, J.B. 1997, *Gaining and sustaining competitive advantage*, Addison-Wesley, Reading (MA).

Bassala, G. 2001, *The Evolution of Technology*, Cambridge University Press, Cambridge.

Berkhout, A.J., P.A. van der Duin, L. Hartmann & J.R. Ortt 2007, *The cyclic nature of innovation: connecting hard sciences with soft values*, Elsevier, Oxford.

Brandenburger, A.M. & B.J. Nalebuff (1995), "The right game: use game theory to shape strategy", *Harvard Business Review*, (July-August), pp. 57-71.

Brown S.L & K.M. Eisenhardt 1997, "The art of continous change: linking complexity theory and time-paced evolution in relentlessly shifting organizations", *Administrative Science Quarterly*, Vol.42, pp. 1-34.

Bush, V. 1945, *Science. The Endless Frontier*, United States Government Printing Office, Washington.

Day, G.S., D.J. Reibstein, et al., eds. 1997, *Wharton on dynamic competitive strategy*, John Wiley & Sons, Inc, New York.

De Geus, A. 1997, *The living company*, Harvard Business School Press, Boston (MA).

Eisenhardt, K.M. & J.A. Martin 2000, "Dynamic capabilities: what are they?", *Strategic Management* Journal, Vol.21, pp. 1105-21.

Epstein, J. & R. Axtell 1996, *Growing artificial societies: social science from the bottom up*, MIT Press, Cambridge (MA).

Hamel, G. & C.K. Prahalad 1994, "Competing for the future." *Harvard Business Review* (July-August), pp. 122-128.

Hannan, M.T. & J. Freeman 1977, "The population ecology of organizations", *American Journal of Sociology*, 82(5), pp.929-64.

Hannan, M.T. & J. Freeman 1989, *Organizational ecology*, Harvard University Press, Cambridge (MA).

Helfat, C.E. & M.A. Peteraf 2003, "The dynamic resource-based view: capability lifecycles", *Strategic Management Journal*, 24, pp. 997-1010.

Johannessen, J.A., J. Olaisen & B. Olsen 1999, Managing and organizing innovation in the knowledge economy, *European Journal of Innovation Management*, Volume 2, Number 3, pp. 116-128.

Levinthal, D.A. 1991, "Organizational adaptation and environmental selection – Interrelated processes of change", *Organization Science*, 3, pp. 47-71.

Liyanage, S. Greenfield, P.F. & Don, R. 1999, Towards a fourth generation R&D management model-research networks in knowledge management, *International Journal of Technology Management*, Vol.18, No.3/4, pp. 372-93.

Miller, W.L. 2001, Innovation for Business Growth, *Research Technology Management*, September-October, pp.26-41.

Nelson, R.R. & S. Winter 1982, *An evolutionary theory of economic change*, Harvard University Press, Cambridge (MA).

Newbert, S.L. 2007, "Empirical research on the resource-based view of the firm: an assessment and suggestions for further research", *Strategic Management Journal*, 28, pp. 121-146.

Ortt, J.R. & P.A. van der Duin. "The evolution of innovation management towards contextual innovation", submitted to the *European Journal of Innovation Management*

Penrose, E.T. 1959, *The theory of the growth of the firm*, Wiley, New York.

Porter, M.E. 1980. *Competitive strategy: techniques for analyzing industries and competitors*, The Free Press, New York.

Porter, M.E. 1985, *Competitive advantage: creating and sustaining superior performance*, The Free Press, New York.

Prahalad, C.K. & G. Hamel 1990, "The Core Competence of the Corporation", *Harvard Business Review* (May-June), pp. 79-91.

Preez, G.T. Du & C.W.I. Pistorius 1999, "Technology threat and opportunity assessment, *Technological Forecasting and Social Change*, Vol.61, Issue 3, pp. 215-234.

Rothwell, R. 1994, "Towards the Fifth-generation Innovation Process", *International Marketing Review*, Vol 11 No 1, pp. 7-31.

Roussel, P.A., Saad, K.M., & Erickson, T.J. 1991. *Third Generation R&D. Managing the link to Corporate Strategy*. Arthur D. Little.

Scherer, F.M. and D. Ross 1990, *Industrial market structure and economic performance*. Houghton Mifflin Company, Boston.

Teece, D.J. et al. 1997, "Dynamic capabilities and strategic management", *Strategic Management Journal*, 18, pp. 509-33.

Twiss, B. 1992, *Forecasting for technologists and engineers. A practical guide for better decisions*, Peter Peregrinus Ltd, London.

Van der Duin, P.A. 2006, *Qualitative futures research for innovation*, Eburon Academic Publishers, Delft.

Van der Heijden, K. 1996, *Scenarios: the art of strategic conversation*, Wiley, Chichester.

Wernerfelt, B. 1984, "A resource-based view of the firm", *Strategic Management Journal*, 5, pp. 171-80.

Yakhlef, A. 2005, "Immobility of tacit knowledge and the displacement of the locus of innovation", *European journal of innovation management*, Vol.8, No.2, pp. 227-2390.

Ybema, S. 2004, "Managerial Postalgia: Projecting a golden future", *Journal of Managerial Psychology*, 19-8, pp. 825-41.

Zegveld, M.A. 2000, *Competing with dual innovation strategies: a framework to analyse the balance between operational value creation and the development of resources*, Werk-veld, Den Haag.

Zegveld, M.A. 2004, *Corporate strategy and the position of technology: a bird's eye view*, in: Reader Technology and Strategy, M. A. Zegveld & E. Den Hartigh, Delft University of Technology, Den Haag, pp. 1-20.

10 Astronomy: observing the past and predicting the future

Dap Hartmann

1 Introduction

Astronomers observe only the past. The light we receive from stars may be thousands of years old. The further away the object is, the further back in time we look. The nearest star to the Sun is Proxima Centauri, at a distance of 4.2 light years.[1] Even though it is the nearest star, we cannot see Proxima Centauri with the naked eye (meaning that we use no binoculars or telescope), because it is too weak. The brightest star in the night sky is Sirius, which lies at a distance of 8.6 light years.[2] The light that we receive today, left Sirius 8.6 years ago. In other words, we observe Sirius today as it was 8.6 years ago – we actually observe the past. Even the light from the Sun is 'old'. The sunlight we see at this moment, left the Sun 8 minutes ago. Strictly speaking, this is true for everything we observe. If you look at a woman across the table, you see her as she was three-billionth of a second ago – the time it took the light to travel the distance to your eyes. Add to this the processing time your brain needs to interpret the images, and you realize that any woman is always slightly older than what you see.

The further we look into the depths of the universe, the further back we look in time. Stars in our Galaxy can be hundreds of thousands of light years away

1 A light year is a *distance* measure, not a measure of time (as sometimes mistakenly assumed, undoubtedly because of the word 'year'). It is the distance that light travels in a vacuum in one year. The speed of light is the 299,792,458 meters per second, which means that a light year is 9.46×10^{15} meters, or roughly 9.5 trillion kilometers.

2 Because stars differ tremendously in their intrinsic brightness, there is no relationship between their apparent brightness in the night sky and their distance. Sirius is 23 times brighter than our Sun, and we can see Sirius with the naked eye. But that same eye located at the distance of Sirius cannot see the Sun. Rigel (in the constellation of Orion) is a whopping 40,000 times brighter than the Sun. But at a distance of about 800 light years, Rigel is only the 7th brightest star in the night sky.

while the nearest galaxy (Andromeda) is 2.6 million light years away. Recently, a massive black hole was discovered at a distance of 12.7 billion light years (Goto, 2006). As the Universe is believed to be 13.7 billion years old, this implies that we are looking 12.7 billion years back in time, and observe this object at the time when the Universe was only 1 billion years old.
Astronomical theories explain the observations ('the past') and make predictions ('the future'). A theory without predictive powers is useless, because it cannot be tested. One of the most stunning examples of the predictive power of a scientific theory, is the accurate prediction of the bending of starlight around the Sun. This was first observed by Eddington during the May 29, 1919 total solar eclipse. According to Einstein's theory of general relativity (Einstein, 1916) light is influenced by the presence of a strong gravitational field. Light from distant stars which passes close to the sun, will be slightly deflected. Eddington traveled to the island of Principe (off the coast of West Africa), where the solar eclipse could be observed at its best (it lasted almost 7 minutes). Unfortunately, the sky was overcast, but during the last 10 seconds it cleared up and Eddington managed to take just one photograph. That picture showed that the position of the background stars (visible because the Sun was obscured) was slightly deflected from the true position of the stars, in accordance with Einstein's prediction. A wonderful triumph for the predictive power of scientific theories, in particular Einstein's theory of general relativity. The beauty is further enhanced by the fact that the confirmation of this prediction was achieved by observing a solar eclipse, which itself was very accurately predicted.

Our everyday modern life is only possible thanks to the predictive power of science. The range of predictions covers an enormous spectrum in space and in time. At any date – past, present or future – we know exactly what the phase of the Moon is, when the tides occur, and to what extent. We can compute when particular constellations of the planets and moons in our solar system will occur. That way we know that Venus passed in front of the Sun on June 8, 2004. The previous time that happened, was 122 years earlier (in 1882), and the next time will be on June 6, 2012 (Britt, 2004). We know that on Aug 28, 2007 a total lunar eclipse will be visible in East Asia, Australia, the Pacific, and the Americas, and that it will last one hour and 31 minutes. The first total lunar Eclipse visible from Europe is on Feb 21, 2008, and will last 51 minutes. Amazingly, astronomers have calculated that there will be an total lunar Eclipse on November 4, in the year 3000. NASA (2007a) contains a catalog of 5000 years (2000 BC-3000 AD) of lunar eclipses. Similarly, NASA (2007b) lists solar eclipses over the same period. Although there may not be many practical applications of eclipses, legend has it that Christopher Columbus intimidated the natives in Jamaica, where he was stranded for a year, by predicting the lunar eclipse of February 29, 1504 with the help of an ephemeris (Morrison, 1955, pp. 184-92).

Astronomy should not be confused with astrology, its linguistically similar ugly cousin. Astrology pretends to be able to predict the future, where, in fact, it can do no such thing. Unfortunately, in the days of Newton and Kepler, astronomy and astrology were rather close scientific cousins. According to Kusukawa (1999): "Kepler believed in astrology in the sense that he was convinced that planetary configurations physically and really affected humans as well as the weather on earth. He strove to unravel how and why that was the case and tried to put astrology on a surer footing, which resulted in *On the more certain foundations of astrology* (1601). In *The Intervening Third Man, or a warning to theologians, physicians and philosophers* (1610), posing as a third man between the two extreme positions for and against astrology, Kepler advocated that a definite relationship between heavenly phenomena and earthly events could be established." Whereas astrology claims to have the power to predict the future, only astronomy is really capable of doing so.

2 Astronomy and Physics

Astronomy is basically physics applied at the very largest scale – the scale of the universe. An essential assumption is that the laws of physics as we have discovered them on Earth, are identical throughout the entire universe. Even though there are many things that we still do not understand about the universe, the laws of physics appear to be valid everywhere.[3] It is bad practice to try and explain some of the unresolved mysteries of the universe by modifying the laws of nature to fit the observations. For example, it appears that a lot of mass is missing from the universe. By observing and analyzing the rotation curves of galaxies, astronomers have calculated the mass required to explain this rotational behavior. However, after adding up all the mass that is visible (over the entire electromagnetic spectrum) only a fraction of the calculated mass is actually accounted for. The missing mass is generally referred to as 'dark matter', suggesting that it must exist but we cannot (yet) see it. This

3 We also assume that the laws of nature are *always* valid. In the words of Richard P. Feynman: "We do not imagine, at the moment, that the laws of physics are somehow changing with time, that they were different in the past than they are at present. Of course, they may be, and the moment we find they are, the historical question of physics will be wrapped up with the rest of the history of the universe, and then the physicist will be talking about the same problems of astronomers, geologists, and biologists" (Feynman, 1963). Observations from distant quasars suggest that the laws of nature may have been slightly different in the very early days of the Universe (Webb, 2003).

is the most commonly accepted explanation. Alternatively, some scientists have suggested that the law of gravity at very large scales[4] might be (slightly) different from the gravitational law that we have established for much smaller scales. If this were true, there simply is no missing mass. However, deviations from the laws of gravity at large scales would lead to other effects, none of which have been observed.

Physics is about explaining the physical world around us by observing how matter and (electromagnetic) waves behave. It appears that most of this behavior is not a random, but obeys very strict rules. These rules, once discovered, tested and formalized, are called of 'laws of physics' or even 'laws of nature'. Because matter and waves obey these laws, their behavior is highly predictable, which enables us to predict their future. For example, Newton's first law of motion states that a mass[5] which moves at a constant velocity will carry on moving in a straight line at that velocity, as long as there is no force acting upon it. In our everyday experience, that condition is not fulfilled, and even in the vast emptiness of interstellar space those idealized conditions never actually occur. Yet, in first approximation, Newton's first law holds perfectly for an object (a space probe, for example) that is far away from the Sun and the heavy outer planets of our solar system. Pioneer 10 was launched in 1972 and has now traveled beyond the orbit of Pluto. It is the first man-made object to leave the solar system. Pioneer 10 is currently about 14 billion kilometers (13 light minutes) away from Earth, and will continue to travel at its present velocity (about 12 km/s, or 44,000 km/h). We can accurately predict where it will be in a hundred years from now. Unfortunately, we have lost radio contact with the spacecraft, which makes it impossible to track its further journey. But we know that Pioneer 10 is heading in the direction of Aldebaran, the brightest star in the constellation of Taurus (and one of the brightest stars in the night sky). If Aldebaran were a stationary object, it would take Pioneer 10 some 1.6 million years to reach it. However, Aldebaran recedes from us with a velocity of 54 km/s (4.5 times faster than Pioneer 10 is approaching), so we know that the spacecraft will never reach it. A nice illustration of this great predictive power of physics can be found at http://www.heavens-above.com/solar-escape.asp, which shows up-to-date information on five spacecraft which are leaving the solar system.

4 The diameter of our galaxy is about 100,000 light years. The Andromeda galaxy, a neighbor of our galaxy, is 2 million light years away. The size of the observable universe is estimated at 13 billion light years.

5 A 'mass' is how physicists refer to an object. Ideally, an object has no physical dimensions, because that makes the equations (calculations) a lot simpler.

Meanwhile, back on planet Earth the situation is quite different. If you throw a baseball straight ahead, it will not travel in a straight line. Newton's first law does not apply because the Earth's gravitational field exerts a force on the baseball which deflects it from a straight line. Luckily, we also understand this process extremely well, and this allows us to calculate what happens to the baseball. Because we are able to calculate this behavior in advance, we can accurately predict what will happen when we actually do the experiment. In other words, we are able to predict the future. If we throw a baseball straight ahead from a 100-m high tower at 100 km/h[6], the baseball will hit the ground after 4.51 seconds at a point which is 125 meter away from the tower. Surprisingly (to people who know little about physics) it would also take 4.51 seconds for a baseball to reach the ground if we just dropped it. The horizontal velocity that we supply to the ball when we throw it will not delay it falling to the ground. When you drop a baseball from a height of 5 meters[7] it will take exactly 1 second to hit the ground. When you drop it from a height of 20 meters, it will take two seconds (not four seconds, as you might have guessed if you do not know much about physics). Drop the baseball from the 100 meter high tower and it hits the ground in 4.51 seconds. Throw it at 100 kilometers per hour and it will also hit the ground after 4.51 seconds. But in this case, the baseball had 4.51 seconds to travel at 100 km/h, and therefore it lands 125 meters away from the tower. The combined trajectory of moving horizontally and falling down is perfectly described by parabola. Long before Newton explained exactly *why* it behaves like this, people already knew that it *did* behave like that. More than 2000 years ago, cannons fired projectiles which followed parabolic trajectories. That had nothing to do with the cannon, which only served as a device that provided the forward thrust. The right combination of initial speed (dependent upon the weight of the projectile and the amount of gunpowder used) and the angle at which the cannon was raised, determined the distance the projectile traveled. Elementary physics allowed gun operators to predict where their cannonballs would land – preferably on an enemy ship or settlement. Shoot straight up, and the cannonball will fall straight down again; shoot straight ahead, and the cannonball lands relatively nearby. To reach the greatest distance, the angle should be 45 degrees.

A theory that explains extremely well and in relatively simple terms how nature works, is often called a physical law or a law of nature. The baseball example above illustrates that once a law of nature has been discovered, it has

6 Professional baseball pitchers can throw almost twice as fast.

7 Actually, it is 4.9033 meter, and this is accurate to the last digit (which amounts to a fraction of a millimeter). The accuracy is so high that it actually depends on the specific location on Earth. The Earth is not a perfect sphere and therefore the exact distance from the surface to the center of the Earth varies.

great predictive powers. Another characteristic of a scientific theory is that it can predict things that were hitherto completely unknown. Confirmation of such predictions strengthens the belief in the validity of the theory. When Mendeleev created his periodic table of atomic elements in 1869, there were a few empty slots in his chart. The periodic table is a systematic way of classifying atomic elements according to certain observed chemical properties (as a result of their physical properties) that they have in common. For example, gaseous elements which do not react with other elements (for example, they do not react with oxygen, which means that they do not 'burn') are all located in the same column of the periodic table. These gasses are known as 'nobel gasses'. On the other hand, alkali metals (such as sodium and potassium) are extremely reactive and are therefore rarely found as pure elements in nature. A little potassium thrown in water bursts into flames, while caesium (another alkali metal) creates a violent explosion (Alkali Metals, 2007). The empty spaces in the periodic table suggested that atomic elements ought to exist with chemical properties according to were they were located in the table. For example, in 1871, Mendeleev predicted the existence of an element he called ekaaluminium (because it occupied the empty slot below aluminum), and for which he estimated an atomic mass of 68 (Mendeleev, 2007). In 1875 it was discovered by the French chemist Lecoq de Boisbaudran. The element which we now call gallium has an atomic weight of 69 (Gallium, 2007).

3 Peril from within the solar system

The Earth is constantly being bombarded with meteoroids (tiny rocks from space). Most of them are sized between a grain of sand and a 10-inch rock, and burn up in the Earth's atmosphere. When that happens, a trail of light is visible which is commonly referred to as a meteor or a shooting star. If a meteoroid is large enough, it will not burn up completely and part of it will strike the Earth's surface. The rock that hits the Earth is called a meteorite. The bigger the meteoroid, the bigger the meteorite and the bigger the impact (crater). One look at the Moon (which has no protective atmosphere) and its craters is enough to get a sense of what it means to constantly be bombarded by rocks from space. More than half a million craters on the Moon can be seen from Earth; the largest are more than 200 km in diameter.

We cannot predict when a big meteoroid will strike the Earth, because we have insufficient data. Cosmic debris is abundant, but we do not know exactly where it is, and how it moves through space. If a space rock is on a collision course with Earth, there are too many uncertainties to make highly accurate predictions. If it is coming straight at us, it is very difficult to measure its velocity perpendicular to the line of approach. It might miss the Earth by 100,000 km, which may seem like a comfortable margin but is, in fact, an extremely narrow miss (it is only one third of the distance to the Moon). If we assumed that it would strike the Earth, and if we were capable of deflecting it from its path, we might actually accomplished exactly the opposite and change a near miss into a head-on collision. Blowing it up into a million smaller pieces is also not a good idea. Even though we have never actually blown up an approaching meteoroid, we can calculate what would happen by simulating it with a powerful computer. The energy contained in the million fragments together is just as big as the energy of the original rock. The Earth will now be hit by buckshot, and the chances of a miss have dramatically decreased. That is why hunters shoot ducks with buckshot and not with a single bullet. Buckshot of meteoroid fragments might just blow away the Earth's atmosphere entirely.

4 The fate of our Sun

Our Sun is only an average star which has been shining for about 4.5 billion years. Incredible as it may sound, we know exactly what will happen to our Sun in the very far future. In about 5 billion years from now, the nuclear processes in the core of the Sun will run out of fuel (hydrogen), and gravity will make the core contract. That sets off nuclear fusion in the outer layers of the Sun, which will cause it to expand and become a so-called red giant (see also footnote 9). The Sun will have expanded beyond the radius of the Earth's orbit, which means that the Earth will literally be vaporized. The core of the Sun eventually becomes a white dwarf, surrounded by a planetary nebula (the remnants of the outer gas layers and the evaporated planets). How on earth is it possible that we know all that?

Astronomers have observed so many stars in all the different stages of their lives, that it paints a detailed picture of how stars actually evolve. Stellar evolution is a highly advanced research field which is capable of making accurate predictions about the lifecycles of stars. It turns out that the initial mass of a star is the dominant factor that determines its further evolution. A heavy star will eventually become a supernova – an incredibly powerful explosion which causes the star to become brighter than the entire galaxy it is contained in. It has been established that in 1054, Chinese and Arab astronomers have witnessed a supernova. An insignificant star in the constellation of Taurus blew up and became so bright that it could be seen during the day. Almost a thousand years later, we can still see the remnant of that supernova. This supernova remnant is known as the Crab Nebula, because of the irregular appearance of the gas that was blown away from the exploding star and its interaction with the surrounding interstellar medium (gas and dust). In the center of this nebula are the remains of the original star: a neutron star which rotates at a speed of 30 revolutions per second and emits powerful pulses of radio waves. Such an object is called a pulsar.

5 Watching history unfold

Looking up at the sky on a dark and clear night, we can see a great many stars. The longer we look, the more stars we see, especially if we are far away from the city. With the naked eye we can see approximately 8000 stars (4000 in the northern hemisphere, and 4000 in the southern hemisphere. At any one location, we can see about 4000 stars with the naked eye). All these stars reside in our galaxy, which is also known as the Milky Way.[8] The total number of stars in our galaxy is around 100 billion, and so with the naked eye we see only a very tiny fraction. For every star that we *can* see, there are another 12 million stars that we *cannot* see without using a telescope. Aldebaran is one of the brightest stars in the night sky, and has a very distinct red color. It lies at the distance of 60 light years from Earth, which means that the light we see today left Aldebaran 60 years ago. What it looks like right now will only become apparent in another 60 years, when the light it emits today will finally have reached us. We understand stellar evolution so well that we can confidently say (predict) that Aldebaran will still be there in 60 years. It will not have blown up, it will not have died out and it will not even look any different – not less red, not weaker, etc. If Aldebaran has a planetary system[9], and if there is intelligent life on one of the planets, then these aliens may look up at the sky and observe our star, the Sun. But unless their eyes are much more sensitive than ours, they would need a telescope to see it, because the Sun is considerably weaker (on an

8 The Milky Way appears as a light band across the sky. Many ancient cultures interpreted it as milk that had been spilled. For example, according to Greek mythology, Heracles was the son of Zeus and Alcmene (a mortal woman). Zeus let his sleeping divine wife Hera breastfeed Heracles. When she woke up she was startled and the milk that spilled became the Milky Way.

9 As of May 2007, 236 exoplanets (extrasolar planets, meaning planets beyond our Solar system) have been discovered (Schneider, 2007). Although a tentative discovery of a planet orbiting Aldebaran was announced in 1997, no confirmation has since been reported. The science-fiction novel *Narabedla Inc.* by Frederik Pohl (1988) is set on a planet circling Aldebaran (Narabedla is Aldebaran spelled backwards). Sol Company (2007) states that: "any Earth-type planets that orbited Aldebaran during its youth would have been burnt to a cinder by now, and possibly fallen into the star from frictional drag with the giant star's gaseous envelope. Astronomers would find it very difficult to detect an Earth-sized planet around this star using present methods." Aldebaran has evolved into a red giant with a diameter of about 50 times that of the Sun.

absolute scale) than Aldebaran. If today (2007) the aliens from Aldebaran look at our Sun, they see it as it was 60 years ago. If they have a very, very powerful telescope, they might see the Earth as it was in 1947. If they possess an impossibly powerful telescope[10], they can actually witness how a massive meteorite hits the Earth near the village of Paseka, 440 km northeast of Vladivostok[11]. In other words, they would be looking today at something that happened here on Earth 60 years ago. To push this thought experiment even further, suppose that the aliens from Aldebaran use a large mirror to reflect those images back at us. If we too had an impossibly powerful telescope, we would be able to see that meteorite strike the Earth in 60 years from now. Today, we would be able to see the premiere of Verdi's opera Otello in Teatro alla Scala in Milan (Italy).[12] It took these images 60 years to reach Aldebaran (where they hit the mirror), and another 60 years to travel all the way back to us. What this means, is that the images of our history are traveling through space at the speed of light. Remember, this is merely a thought experiment and it is not practically possible to carry out what I discuss here. Nevertheless, theoretically these are true facts, even though we cannot put them into practical use. It is not unlike the philosophical argument that 'when you spit in the sea, the sea will never be the same again'. Or that with every breath you take, you inhale a few molecules of Julius Caesar's last breath. Theoretically, images of dinosaurs roaming the Earth are forever traveling through space.[13] Far-fetched as this may seem, there are two more realistic examples that I would like to mention: one is based on facts; the other is fictional (although entirely possible).

The first example is the message that we (the human race) have deliberately sent out into space. Radio signals are just another instance of electromagnetic radiation. We are very familiar with light (which is electromagnetic radiation with a wavelength between 400 nm and 700 nm[14]), because we possess two sensors to detect it: our eyes. Radio waves, X-rays, microwaves, and infrared radiation are different manifestations of electromagnetic radiation, but at different wavelengths. All electromagnetic radiation propagates at the same

10 'Impossible' here means that it is considered to be *practically* impossible, while theoretically it might still be possible.

11 This actually happened on February 12, 1947 (Sikhote-Alin_Meteorite, 2007). The weight of the meteorite before entering the atmosphere is estimates at 900,000 kg. About 100,000 kg reached the surface of the Earth.

12 Otello premiered on February 8, 1887.

13 Dinosaurs roamed the Earth between 230 million and 65 million years ago. It is believed that they became extinct after a massive meteorite hit the Earth. The images of this catastrophic event have now travelled to a distance of 65 million light years from Earth. That means that aliens in the Virgo Cluster (a concentrated group of some 1300-2000 galaxies) with an impossibly powerful telescope can watch that meteor strike the Earth today.

14 A nanometer (nm) is one billionth (10^{-9}) of a meter. Red light has a wavelength of about 700 nm, which means that more than 1400 full waves of red light fit into one millimeter.

velocity (300,000 km/s) through space. In 1974, astronomers used the biggest radio telescope on Earth (the Arecibo radio telescope in Puerto Rico, which has a diameter of 305 meter) to send out a message into space. This message, which is now known as the 'Arecibo message', consists of a graphical image (in black and white, with a typical 1970s low resolution of only 23 by 73 pixels) containing information about our location in the solar system (third rock from the Sun), what we look like (sticklike figures), and a double helix structure (indicating that we have figured out where our genetic code resides). There is more information in this picture, but even the aforementioned data are not all that obvious.[15] It is clearly aimed at a technologically advanced civilization that can receive our message, recognize it as a message, decode it properly[16], and interpret it. A powerful transmitter and a collimated beam were used to send out this signal in the direction of M13, a globular cluster at a distance of about 25,000 light years. M13 is a large and dense cluster of stars that happened to pass over the Arecibo telescope at the time.[17] Not everyone was thrilled about this experiment. Undoubtedly fueled by the many science-fiction stories which depict aliens as evil warmongers, there were concerns about this reckless 'Hello, we are over *here*!' message sent out into space.

Figure 1 The Arecibo Message. A 23x73 pixel binary image was transmitted into outer space on November 16, 1974, using the largest single-dish radio telescope on earth (the 305-m Arecibo radio telescope). If you (as a human interpreter) cannot figure out what the message is, take a look at Arecibo_message (2007), and remember that it is intended for extraterrestrial civilizations.

15 Figure 1 shows the Arecibo message. Try to decode this black-and-white image; remember, you are looking at the already decoded image and you are human (just like the people who designed the message). If you cannot figure it out, look up Arecibo_message (2007).

16 When the 23x73 image is displayed as 73x23 it looks like garbage. But actually, the resolution was cleverly chosen: 23 and 73 are prime numbers (see footnote 19). The product (1679) is the number of bits of information contained in the message, which can only translate into a rectangular grid as 23 columns and 73 rows (or 73 columns and 23 rows).

17 The Arecibo telescope cannot be pointed in an arbitrary direction. It is stationary and sees only what happens to pass overhead.

The second example has a close link to the story above. It occurs in the science-fiction novel *Contact* by Carl Sagan (1985). Sagan (1934-1996) was an astronomer specialized in exobiology (the study of life in space, combining knowledge from astronomy, biology and geology), and was closely involved with the Arecibo message experiment. Science fiction may be regarded as a playing ground for extrapolating our current scientific knowledge and technological capabilities, and philosophizing about the consequences. In fact, several distinguished scientists have written science fiction novels, including astronomers Carl Sagan, Fred Hoyle and David Brin, and physicists Gregory Benford and Robert L. Forward. In *Contact*, an artificial signal is received by a female astronomer.[18] Electromagnetic radiation from natural processes does not have regularities (patterns). When a signal is received that does contain a pattern, it can mean three things: (1) it was man-made and is accidentally observed by the telescope as if coming from space; (2) it is alien-made, in which case most religions on Earth need serious revisions; or (3) it is caused by an unfamiliar physical process. The latter actually happened in 1967, when an extremely regular pulsating signal (at a rate of one every few seconds) was received from the depths of space. It was ruled out as being not man-made because the signal reappeared after one sidereal day, not after one solar day. There was no known natural process that could explain this astronomical beacon, and so the astronomers joking-ly named it LGM-1 (Little Green Men, number one). It turned out to be a very fast rotating neutron star. The extreme regularity of the pulse we receive, corresponds to the rotationial period of the neutron star. These objects are called pulsars.

In *Contact*, when a signal from outer space is received, it is ruled out as man-made and also as a pulsar. That leaves just one interpretation: it is of alien origin – possibly a message. Note that not every signal (alien or man-made) contains a message. But in *Contact*, it was indeed a message. The aliens used two clever tricks to help the humans decode their message. The first trick was to use a sequence of prime numbers[19] at the beginning of the message. Mathematics is believed to be universally valid, which means that intelligent beings elsewhere in the universe have identified the same prime numbers as we have. Receiving a sequence of prime numbers implies that an intel-

18 The astronomer is clearly modeled after Jill Tarter, one of the key scientists in the Search for Extra Terrestrial Intelligence (SETI) program. The SETI project searches the sky for messages from other worlds.

19 Prime numbers cannot be divided by a number other than one and the prime number itself. There are infinitely many prime numbers, starting with the sequence 2, 3, 5, 7, 11, 13, 17, 19, 23. No formula exists to calculate prime numbers; there is no pattern that prime numbers conform to.

ligent being sent a message, because there is no known natural (physical) process that generates prime numbers. Prime numbers are unmistakably the signature of intelligent beings. Because there is obviously no way for the aliens to know when we would receive their message, it was repeated indefinitely. This also helped the humans to determine when the entire message had been received. The second trick was to use a familiar carrier signal in which the real message is contained. The question is: what carrier signal could the aliens use that is familiar to the inhabitants of planet Earth? Very cleverly, they used a signal that they had first received from us! Did we ever send out a message into space, prior to the Arecibo message? Not deliberately, but just as light from Earth (sunlight reflected off the Earth) travels through space, so too do man-made radio waves. What the aliens intercepted, was the live TV broadcast of the opening by Adolf Hitler of the 1936 Olympic Games in Berlin. Radio waves containing these TV images left the Earth in 1936 and had been received by the aliens living on a planet orbiting the star Vega.[20] To help us decode their message, the aliens sent back the same images and interleaved them with their actual message. In other words, it was as if there was a mirror on the alien planet which reflected the 1936 TV broadcast. This also sets the maximum distance to the aliens, namely the distance that light can travel in half the time since the original broadcast. If we were to receive those 1936 TV images in 2016, the aliens are at most 40 light years away from Earth.

Physics (as we understand it today) prohibits anything from traveling at a speed greater than the speed of light in a vacuum (299,792.458 km/s – 300,000 kilometers per second as a fair approximation). This not only pertains to physical objects, but also to less tangible things such as 'information'. The transfer of information needs a physical mechanism. When you shout, sound waves propagate through the air (or throught water, where the speed of sound is actually higher than in air). When you use a flashlight to transmit a Morse code, light waves carry that message. There is absolutely no way to send a signal containing information faster than the speed of light. The implication of this physical limitation is that there are events which are fundamentally outside of our reach. We cannot influence these events in any way, because we cannot get there in time – not physically and not by sending a message. Suppose, for example, that we have sent an astronaut to the planet Mars. There are serious plans to actually do this, and it might happen within the next 25 years. The distance to Mars varies between 60 million km and 400 million km, depending on the relative positions of the Earth and Mars in their

20 In the night sky, only four stars appear brighter than Vega. The distance to Vega is about 25 light years.

orbits around the Sun. Let us assume that when the astronaut is on Mars, the distance is 270 million km, which is exactly 15 light minutes. Suppose that immediately after the landing, the astronaut notices that his oxygen tank has sprung a leak. If he cannot fix the leak, all the oxygen will escape in 25 minutes and he will suffocate to death. He contacts mission control, asking the technicians on Earth what to do. As he is 15 light minutes away from the Earth, it will be 15 minutes before Mission control receives his distress call. Even if the technicians on Earth responded immediately, the answer still takes another 15 minutes to get back to the astronaut. If he has not figured out how to fix the problem himself, he will be dead before the answer reaches him. There is simply nothing anyone on earth can do to interfere with what is happening on Mars, on a timescale shorter than the time it takes light to travel that distance. Figure 2 illustrates this inescapable law of nature graphically.

Because we cannot draw a four-dimensional picture, the three spatial dimensions have been reduced to two (representing space as a plane that is perpendicular to the page). The time axis runs vertically. The center point (indicated as 'observer') represents 'here and now'. The cone that extends upwards is created by a single flash of light at time='now'. In reality, such a light flash propagates spherically in three dimensions. The sphere gets bigger with time; the 'wave front' moves through space at the speed of light. In the two dimensions of figure 2, the wave front takes the shape of a circle. After one second, the radius of that circle measures one light second (or 300,000 km). It is like throwing a stone in a pond: after one second there is a circle in the water with a radius equal to the distance the water waves have traveled in that one second. In Figure 2, the time axis runs vertically, so all the light circles stack up and create a cone. Only the space enclosed by that cone can be reached by light departing from the present; everything outside the cone can not be reached. That is one of the consequences of Einstein's theory of relativity; this is how the Universe works.

Note that in the absence of spatial motion, the 'here and now' remains 'here', while the 'now' moves up as time increases. That means that we can stay fixed in space[21], but we cannot stay fixed in time. In other words, the future is inevitable – inescapable. Also, we can move back and forth in space at an arbitrary rate (not exceeding the speed of light), but we cannot move back and forth in time. We can only move forward, at a pace that we cannot influence.

21 Relatively speaking, of course. There exists no absolute reference frame in space, but if we consider everything relative to our position on Earth, the argument holds.

Figure 2 Two light cones extending from the 'here and now' (indicated by 'observer'). For the sake of simplicity, the three spatial dimensions have been reduced to two in this image, making up a plane that is perpendicular to the page. Time is the fourth dimension of the space-time continuum, and runs vertically. Upwards from the observer ('now') lies the future, downwards is the past. The 'hypersurface of the present' represents the entire Universe at a single point in time ('now'). The cone that extends upwards represents light which radiates in all directions from the 'now' point in time. Nothing can exceed the speed of light, and therefore this light cone describes the limit of the influence that the observer can exert on the future. There is no way in which the observer can influence space-time outside the light cone. Similarly, the cone that extends downwards defines the boundary beyond which nothing could have influenced the 'here and now'. (Illustration by courtesy of K. Aainsqatsi)

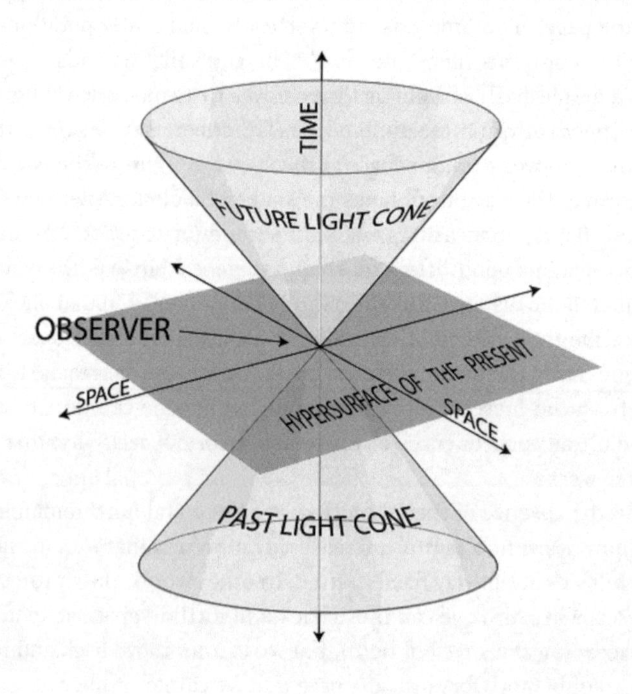

Because the planet Mars (and the unfortunate astronaut on it) lie outside of the light cone (outside of the circle around the point in time where the oxygen runs out), we cannot reach the astronaut by any means. Everything outside the light cones is beyond our direct influence.
The cone extending downwards from the 'here and now' represents the volume of space-time that could have exerted an influence on the present. Only what lies inside this cone could have influenced the observer in the here and now. Light cones demarcate the maximum extent (in space and in time) of the past and the future relevant to what happens in the present.

6 Conclusion

Every astronomical observation contains only information about the past. Even the light from the nearest star (our Sun) is already eight minutes old by the time it reaches the Earth. Light received from the furthest known objects in the Universe has traveled billions of years through space before our telescopes observe it. By analyzing all these 'old data' and developing theories that explain what we observe, astronomers have created very powerful instruments to accurately predict future events. We know exactly when the next total solar eclipse takes place: on August 1, 2008 at 9:24 UT in Grise Fiord, Canada; at 10:20 UT in Nadym in Russia; and at 11:15 UT in Jiquan in China. We know that Halley's Comet will return in July 2061 (on July 29, 2061, it will be closest to the Earth, at merely half the distance to the Sun). And we know fairly well what will happen to our Sun over the next few billion years. In about 5 billion years from now, the hydrogen that fuels the nuclear fusion processes in the core of the Sun will run out, and the Sun will grow bigger and bigger, until it extends beyond the Earth's orbit. That means that the Sun will literally swallow up the Earth. Astronomers can really predict the future, and the future is that we are doomed. That is, if human beings still exist at that time, and if we haven't moved elsewhere in the galaxy. But don't worry about it too much – we have 5000 million years to come up with a cunning plan.

References

Aldebaran, 2007, http://en.wikipedia.org/wiki/Aldebaran (visited 2007.04.03)

Alkaline Metals, 2007, http://www.youtube.com/watch?v=QSZ-3wScePM (visited 2007.08.22)

Arecibo_message, 2007, www.physics.utah.edu/~cassiday/p1080/lec06.html (visited 2007.04.10)

Einstein, A., 1916, Die Grundlage der allgemeinen Relativitätstheorie, *Annalen der* Physik, 49, 769-822

Feynman, R.P., 1963, The Feynman Lectures on Physics, Vol.1, Addison Wesley

Gallium, 2007, http://en.wikipedia.org/wiki/Gallium (visited 2007.08.22)

Goto, T., 2006, http://www.spaceref.com/news/viewpr.html?pid=20715 (visited 2007.06.04)

Kusukawa, S., 1999, http://www.hps.cam.ac.uk/starry/keplerastrol.html (visited 2007.06.11)

Mendeleev,2007,http://en.wikipedia.org/wiki/Mendeleev's_predicted_elements (visited 2007.08.22)

Morrison, S.E., 1955, *Christopher Columbus, Mariner*, Little and Brown, Boston.

NASA, 2007a, http://sunearth.gsfc.nasa.gov/eclipse/LEcat/LEcatalog.html (visited 2007.06.11)

NASA, 2007b, http://sunearth.gsfc.nasa.gov/eclipse/SEcat5/catalog.html (visited 2007.06.11)

Pohl, F., 1988, *Narabedla, Ltd.*, Ballantine, Del Rey, ISBN 0345360265

Sagan, C., 1985, *Contact*, Simon & Schuster, New York

Schneider, J., 2007, *Interactive Extra-solar Planets Catalog*, http://exoplanet.eu/ (visited 2007.05.27)

Sikhote-Alin_Meteorite, 2007, http://en.wikipedia.org/wiki/Sikhote-Alin_Meteorite

Sol Company, 2007, http://www.solstation.com/stars2/aldebaran.htm (visited 2007.05.27)

Britt, R.R., 2004, http://www.space.com/scienceastronomy/venus_transit_040608.html (visited 2007.06.11)

Webb, J., 2003, Physics Word, April 2003, pp.33-38

11 Future-oriented Technology Analyses: The literature and its disciplines

Alan L. Porter

1 Introduction

"Who's doing what" to advance the field of Future-oriented Technology Analyses (FTA)? This paper draws on two motivations. One was to provide background characterization in conjunction with the Second FTA Seminar (Seville, 2006). The second was to provide perspective for this volume.[1] Characterization of the disciplines publishing in this area helps to frame their differing perspectives on futures research.

Beginning about 2000, a number of us have been working to nurture development of FTA. We prepared a perspective paper[2] to help instigate an initiative to share knowledge and promote new methods. This reviewed the various forms of such future-oriented analyses and their history. It noted the marked shift from relatively narrow focus in the earlier years on technological systems to broadened considerations attendant to innovation in socio-economic contexts. Advances in complexity sciences offer new approaches to deal with such systems. The paper noted the increasing attention to science-based, as opposed to technology-based, innovations. It pointed to the increased availability of science and technology information resources in electronic form as enabling potent new analytics. And it compared interests among government, industry, and academia.

1 The former was generated in the context of the Technical Committee for the Second FTA Seminar, of which I served as a member. The second arose in personal discussion with Patrick van der Duin.

2 Coates, V., Faroque, M., Klavins, R., Lapid, K., Linstone, H.A., Pistorius, C., and Porter, A.L., On The Future of Technological Forecasting, *Technological Forecasting and Social Change*, Vol.67, No. 1, p. 1-17, 2001.

The next phase in the FTA progression involved an expanded group preparing a baseline paper for the First FTA Seminar [Seville, 2004; organized and hosted by IPTS – The Institute for Prospective Technological Studies (www.jrc.es)]. This paper[3] explicitly noted analytical forms that address future prospects, relating especially to technology. These include:

- Technology monitoring, technology watch, technology alerts (gathering and interpreting information)
- Technical intelligence and competitive intelligence (converting that information into usable intelligence)
- Technology forecasting (anticipating the direction and pace of changes)
- Technology roadmapping (relating anticipated advances in technologies and products to generate plans)
- Technology assessment, and forms of impact assessment, including strategic environmental assessment (anticipating the unintended, indirect, and delayed effects of technological changes)
- Technology foresight, also national and regional foresight (effecting development strategy, often involving participatory mechanisms)

The First FTA Seminar focused on new methods to address the expanded interests and challenges to better inform science, technology, and innovation (*ST&I*) management and policy processes.

IPTS, under the leadership of Fabiana Scapolo, has carried forward the FTA initiative. In 2006, the Second FTA Seminar keyed on experiences and issues in having FTA affect decision-making. A Technical Committee helped formulate conference themes, prepared five background papers, and drew papers from the Seminar to form a book and special issues of *Technological Forecasting & Social Change*, and *Technology Analysis and Strategic Management*. In addition, a special issue of the *International Journal of Foresight and Innovation Policy* is in preparation. At the seminar, working groups tackled a number of overarching FTA issues, generally oriented toward community-building. An ongoing web portal carries these discussions forward (see http://forera.jrc.es/fta/documents.html).

3 Technology Futures Analysis Methods Working Group, [Alan L. Porter, Brad Ashton, Guenter Clar, Joseph F. Coates, Kerstin Cuhls, Scott W. Cunningham, Ken Ducatel, Patrick van der Duin, Luke Georghiou, Ted Gordon, Hal Linstone, Vincent Marchau, Gilda Massari, Ian Miles, Mary Mogee, Ahti Salo, Fabiana Scapolo, Ruud Smits, and Wil Thissen], Technology Futures Analysis: Toward Integration of the Field and New Methods, *Technological Forecasting and Social Change*, Vol. 71, 287-303, 2004.

This paper offers a literature profile[4] on the FTA domain. The profile helps to characterize the growing body of FTA knowledge. It provides a base for this volume's consideration of disciplinary differences in treatment of futures research, especially between FTA and market research orientations.

2 Methods

I downloaded search results from the "Web of Science" on October 16, 2006.[5] I repeated the process on October 30 to capture cited references as well. This seems like a reasonable information resource to use in that it provides excellent coverage of journal articles in the sciences (SCI), decent coverage of engineering, good coverage of social sciences (SSCI), and additional treatment of humanities (AHCI). These sources include technology management and research policy journals. This certainly does not capture all FTA work, no less the broader spectrum that "futures research" can encompass. In particular it leaves out journals not covered by Web of Science (resulting in a bias toward higher prestige and English language, and against developing country literatures). It does not include conference papers (key in certain fields). And, especially vital in the humanities and some social sciences, it leaves out books.[6]

I searched on some 25 different terms, in various combinations. [Some searches failed – e.g., those, such as "future studies" including the term "studies" don't work because this is a stopword; "scenario" and "trend analysis" proved too general.] Terms yielding 10 or more hits are noted in Table 1 [not all phrasing details are shown – e.g., backcasting was also checked as hyphenated "back-casting"]. The premise is that a number of pertinent articles will not use the general FTA terms (e.g., technology foresight or forecasting). For this reason, Delphi, TRIZ, backcasting, and cross-impact were searched explicitly.

4 Porter, A.L., Kongthon, A., Lu, J-C., "Research Profiling: Improving the Literature Review," *Scientometrics*, Vol. 53, p. 351-370, 2002.

5 ISI "Web of Knowledge" website – for information about access, see: http://scientific.thomson.com/index.html

6 Hicks, D., The Dangers of Partial Bibliometric Evaluation in the Social Sciences, *Economia Politica*, Vol. 23, No. 2, p. 145-162, 2006.

Table 1 Individual FTA Term Search Tallies in Web of Science [1996-2006]

Term	Tally
Delphi	2049
technology assessment	1354
(forecast OR forecasting OR forecasts) SAME (technology OR technologies))	334
('technology roadmap' or 'technology roadmaps' or 'technology roadmapping')	242
technology foresight	76
roadmapping	62
Tech intelligence	61
(analysis SAME technologies SAME emerging)	57
futures research	47
TRIZ	47
Backcasting	46
cross-impact	35
('foresight program' or 'foresight programme')	25
technology monitoring	15
technology watch	13
national foresight	12
technology SAME prospecting	12
Tech mining	10

Unfortunately, the two predominant phrases – Delphi and technology assessment – proved highly problematic. Delphi captured certain physics research as well as articles mentioning the company of that name. Technology assessment mainly yielded "health technology assessment," its own arena of detailed evaluations of medical technologies and programs. To deal with these issues, all results were combined in *VantagePoint*.[7] Duplicates were removed. An iterative process was then used to remove irrelevant articles; create a cleaned file; then check further. Different fields were searched (titles, key terms, journals, full abstract records) to exclude records that seemed to address particle physics, neurology or clinical medicine (health technology assessment), and semiconductor processing. Sets of "out-take" records were reexamined for presence of FTA indicator terms. These were individually checked to re-include if they were judged to relate to FTA. The inclusion scope included decision support approaches. The process is imperfect, but it gains some number of Delphi and technology assessment papers that would otherwise be missed.

7 Text mining software especially developed to help clean, analyze, and report on science and technology data search results – see: //www.theVantagePoint.com.

The resultant set contains articles that are quite inclusive of FTA work, yet do not contain too many "non-FTA" pieces. In information science terms, recall is somewhat selective; precision appears to be reasonably strong.
This yields *1018 FTA-related papers* for 1996-2006 (through Oct. 16). This total is close to that were we to sum the individual searches without including "Delphi" and "technology assessment," although those contribute significantly to this final set. Specifically, "Delphi" contributes 177 and "technology assessment" 169 articles. The next section presents basic activity profile results.[8]

3 Basic results: FTA Publications

The first question – how much FTA research publication is there? Figure 1 shows the trend. From 1996 through 2003, this is essentially flat – at a modest level of about 100 articles per year. [Of course this reflects the scholarly literature; it does not capture foresight reports, etc.] Since 2004, coincident with the first FTA seminar, activity seems to be increasing. Note that the value for 2006 is very arguable – the actual count for the year to date is 89; this is herein doubled to approximate what the complete 2006 publications might be (reflecting that the search covers only a partial year and that indexing of articles lags). "Trend analysis" is somewhat encouraging for FTA.

Figure 1 Trend in FTA Publication [Articles relating to Future-oriented Technology Analysis appearing in Web of Science]

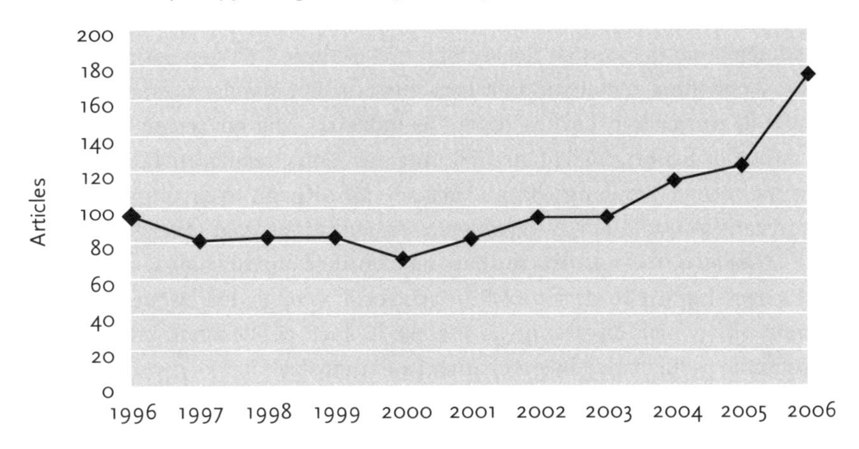

8 Tallies are based on the Oct. 16 searches. The Oct. 30 search counts are very slightly higher. Those search results were matched against the screened results from Oct. 16, to add "cited references."

Figure 2 shows the top 19 (of 55 total) countries (those with 10 or more article authorships). These are extracted from the authors' affiliation and address field. The leading FTA publishing nation is the USA. Given the relative inattention of the US Government to formal technology foresight or assessment, this is rather interesting.

Figure 2 FTA-related Articles (with one or more author affiliations in that country)

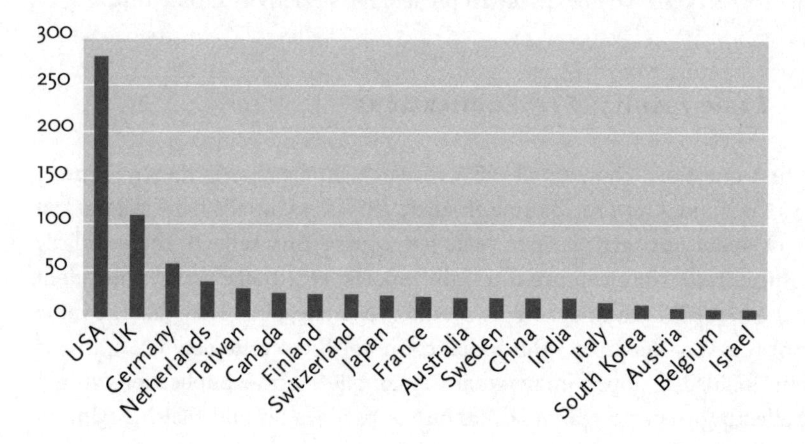

At the institutional level, FTA research appears broadly dispersed; Table 2 lists the top 20 organizations (plus ties). There are 209 organizations with at least two publications.

It is interesting to examine the sectoral mix of these FTA-generating institutions. Applying a thesaurus that combines certain standard phrasing (e.g., "Univ" as university; "Ltd" or "Corp" as industry), and not trying to capture every one of the originally listed 799 organizations, results in Table 3. Note that the second grouping consolidates several difficult to distinguish types – governmental and non-governmental organizations, and other such institutes. [Not all of the 1018 articles have an identified organizational affiliation.] Not surprisingly, publication of FTA articles is strongly led by the academic community (which has the greatest stake in such publication), but note the substantial participation by government and industry.

Table 2 Leading Organizational Affiliations

Affiliation (Name Only)	#
Office of Naval Research (US)	12
Delft Univ Technol	11
Univ Cambridge	11
Natl Chiao Tung Univ	10
Royal Inst Technol	10
Helsinki Univ Technol	9
Univ Illinois	9
Univ Manchester	9
Univ Texas	9
Chalmers Univ Technol	8
Fraunhofer Inst Syst & Innovat Res	8
Swiss Fed Inst Technol	8
Indian Inst Technol	7
Penn State Univ	7
Univ Karlsruhe	7
Univ Penn	7
Georgia Inst Technol	6
Harvard Univ	6
Univ Minnesota	6
Univ Twente	6
Univ Utrecht	6

Table 3 Leading Authoring Organizations by Sector

Type	# of Articles	# of Authorships	% of Articles
Academic	567	779	58%
Gov't/NGO's/Institutes	174	210	18%
Industry	109	142	11%

Table 4 shows where FTA work is being published (11 journals with 10 or more publications). "TF&SC" is the leader, with strong representation of leading technology management journals. Certain special foci are represented. For instance, the "Journal of Cleaner Production" keys on sustainable development, while "Solid State Technology" shows a number of technology roadmapping articles.

Table 4 Leading FTA Journals

Journal	#
Technological Forecasting & Social Change	114
International Journal of Technology Management	52
Futures	49
Research – Technology Management	26
Abstracts of Papers, American Chemical Society	14
Technovation	13
Journal of Cleaner Production	12
Journal of Forecasting	12
R & D Management	11
Solid State Technology	11
Technology Analysis & Strategic Management	11

I merged three fields that provide good topical content: keywords (authors), with keywords (plus), with title NLP (natural language processing) noun phrases.[9] Table 5 gives the flavor of the FTA articles' content via the top 36 key terms (occurring 10 or more times).

Table 5 Leading Key Terms in the FTA Articles

Key Terms	#
technology assessment	92
technology	48
innovation	43
management	36
future	35
science	33
Delphi	32
forecasting	28
Delphi method	26
Delphi study	25
technology foresight	25
Model	22
TRIZ	21
foresight	19

9 Such "text mining" is accomplished using *VantagePoint* software.

sustainability	19
systems	19
models	17
sustainable development	17
backcasting	16
technology forecasting	15
Delphi technique	14
design	14
Impact	13
assessment	12
information	12
bibliometrics	11
energy	11
Industry	11
policy	11
roadmaps	11
Time-series	11
trends	11
application	10
case study	10
roadmapping	10
technical intelligence	10

Figure 3 consolidates variations on topical term themes (e.g., Delphi) to show the trends in usage over this period (number of articles that include the given terms in keywords or title phrases). Delphi, in the context of FTA or relevant decision support, is the most frequently occurring of these forms. Technology Assessment (TA) has declined markedly, although there is a spike of renewed activity in 2005 (not seemingly due to a single concentrated activity). Foresight is younger in vintage than TA. The Second FTA Seminar focused on policy utility of FTA activities; it is somewhat reassuring to see significant article attention to policy aspects of one sort or another. Sustainability and roadmapping both show increasing tendencies, although not uniformly.

Figure 3 Topical Trends in FTA (tallies for 2006 are doubled to normalize this partial year)

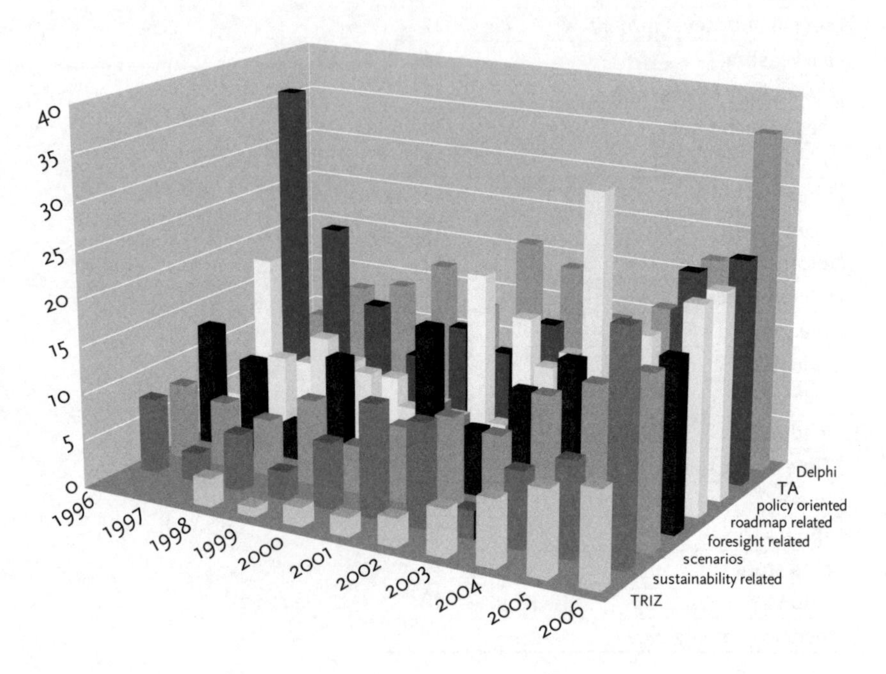

For these analyses, I then expanded the key FTA term selection to include records with these terms in any of the following fields: abstract or title phrases (text-mined via Natural Language Processing – NLP – using *VantagePoint*), keywords (author's), or Keywords Plus (deriving from titles of articles cited by the article in question). I also added "scenarios" to the set included in Figure 3. Table 6 shows the frequency and percentage (of 1018 total articles) mentioning the given term in any of the indicated fields. Many of the terms co-occur in given articles. Overall, at least one of the terms in Table 6 occurs in 67% of the 1018 articles. [Conversely, 33% of the records do not contain any of these terms.]

Table 6 Prevalence of Major FTA Topical Themes and Methods

FTA Topics	# Records	% of Records Mentioning
Delphi	177	17%
TA	169	17%
policy oriented	133	13%
roadmap related	127	12%
foresight related	111	11%
Scenarios	103	10%
sustainability related	83	8%
TRIZ	38	4%

4 Disciplinary contributions to the FTA Literature

Our focal interest is on which disciplines contribute, in what ways, to this literature. Web of Science associates journals with "Subject Categories." About 40% of the journals are linked to more than one of 244 Subject Categories (research areas). This analysis will use these Subject Categories ("*SCs*") as its essential building block. SCs provide a good operationalization of "field" or "discipline" on a fine-grained level.

The author has been involved as an evaluator in a major ongoing project called the National Academies *Keck Futures Initiative* (NAKFI). This is a $40 million, 15-year program to boost interdisciplinary research in the U.S. [www.keckfutures.org]. NAKFI has an interest in measuring the extent of interdisciplinary research, but suitable indicators of such pose a challenge.[10] My colleagues and I have devised measures of research "Integration" and "Specialization" using the SCs.[11] I borrow the underlying tools for these analyses, along with the "domain" categorization of SCs that we are developing for NAKFI.

FTA papers are very widely distributed. The 1018 papers appear in 146 (of the 244 total) Subject Categories, of which 114 show at least two papers.[12] Table 7 shows the top 16 areas (the 17th area shows a sizable drop to 25 articles). This sample of FTA articles appears strongly associated with planning, business, and engineering.

10 Morillo, F., Bordons, M., and Gomez, I. (2001), An Approach to Interdisciplinarity through Bibliometric Indicators, *Scientometrics*, v51n1, p. 203-222.

11 Porter, A.L., Cohen, A.S., Roessner, J.D., and Perreault, M., Measuring Researcher Interdisciplinarity, *Scientometrics*, to appear.

12 The Thomson Scientific Institute for Scientific Information (ISI) associates journals with SCs. A given journal included in the Web of Science could be associated with 1-6 SCs; some 39% of journals in a national sample we studied were linked to more than 1 SC. We thank ISI for providing their thesaurus that links journals to SCs to facilitate these analyses.

Table 7 Leading Subject Categories in which FTA Publications Appear

Subject Category	#
Planning & Development	189
Business	183
Management	174
Operations Research & Management Science	98
Engineering, Industrial	91
Engineering, Multidisciplinary	74
Engineering, Electrical & Electronic	71
Economics	69
Environmental Sciences	61
Energy & Fuels	51
Engineering, Environmental	37
Information Science & Library Science	35
Multidisciplinary Sciences	35
Environmental Studies	34
Engineering, Manufacturing	31
Physics, Applied	31

One way to study research knowledge flows is by using citation patterns. The 1018 FTA papers in this set cite (reference) 21,342 sources (i.e., averaging 21 references per paper). A majority of those cite articles appearing in journals. These 1018 papers cite some 11,793 different sources. We apply a Find-and-Replace thesaurus to standardize how journal title terms are used (e.g., whether J or Jour or Journal). Then we apply a thesaurus to associate journal to SC. In this way we have data on the SCs cited by each paper. This allows us to note what Subject Categories contribute to the intellectual content of these FTA papers. [Obviously, such citation is an imperfect measure as authors reference their own work rather heavily and cite the work of others for multiple reasons.]

In general, the fields in which FTA articles appear most heavily are also those upon which their authors draw, especially planning, business, and engineering. However, two of the leading 16 publication forums (SCs) do not appear as leading sources of FTA intellectual capital – "Engineering, Manufacturing"; and "Physics, Applied." Table 8 shows the most frequently cited Subject Categories. Notice the last two listed. Articles in journals associated with "Political Science" are referenced (cited) by 21 of the 1018 FTA papers. Some of those papers reference a Political Science SC more than once; that is why

the number of instances is higher (45). Thermodynamics is cited by one fewer paper (20), but many more times in those papers (139). For present purposes, I order the cited SCs by the number of FTA articles that mention them.
Table 8 is presented to allow some quick comparisons. In particular note the source areas, upon which FTA authors draw, that are not prominent venues for publication of FTA work (#Pubs). These include several Computer Science SCs, Statistics and Math-oriented SCs, and several Social Sciences.

Table 8 Leading Subject Categories which FTA Publications Cite

Cited Scs	# Records Cited	# Instances Of Being Cited	# Pubs (Top 16)
Management	302	1482	174
Planning & Development	240	686	189
Economics	236	783	69
Business	201	840	183
Operations Research & Management Science	194	546	98
Engineering, Industrial	134	366	91
Multidisciplinary Sciences	125	250	35
Environmental Sciences	110	379	61
Computer Science, Information Systems	96	254	
Information Science & Library Science	93	332	35
Engineering, Multidisciplinary	76	112	74
Engineering, Electrical & Electronic	67	215	71
Social Sciences, Mathematical Methods	64	99	
Computer Science, Interdisciplinary Applications	61	148	
Energy & Fuels	61	347	51
Statistics & Probability	59	131	
Environmental Studies	57	175	61
Computer Science, Theory & Methods	56	135	
Mathematics, Interdisciplinary Applications	55	85	
Computer Science, Artificial Intelligence	51	115	
Social Sciences, Interdisciplinary	51	68	
Social Issues	50	67	
Communication	42	61	
Computer Science, Hardware & Architecture	39	72	
Computer Science, Software Engineering	39	94	
Ecology	38	137	

Cited Scs	# Records Cited	# Instances Of Being Cited	# Pubs (Top 16)
Sociology	38	82	
Mathematics, Applied	37	84	
Engineering, Environmental	36	91	37
Meteorology & Atmospheric Sciences	36	140	
Engineering, Civil	32	48	
History & Philosophy Of Science	31	47	
Public Administration	31	58	
Engineering, Chemical	30	113	
Mathematics	30	32	
Water Resources	30	81	
Psychology, Multidisciplinary	28	36	
Engineering, Manufacturing	27	72	
Telecommunications	27	68	
Education & Educational Research	26	40	
Geosciences, Multidisciplinary	24	68	
Business, Finance	23	99	
Psychology, Applied	22	34	
Political Science	21	45	
Thermodynamics	20	139	

5 Domains

To grasp patterns, it helps to consolidate the 244 SCs into mega-disciplines, or "*research domains*" (in part because they do not neatly reflect disciplinary groupings). I composed these starting with the grouping of SCs into 9 larger sets offered by Morillo.[13] I refined this based on perceptions of commonalities and general publishing practices. To check these clusters, I examined the co-citation of SCs reflected in a combination of four weeks of US-authored publications from 2005 and 2006 in Web of Science (28,922 articles). I examined several alternative ways to group (cluster) the SCs:

- Two Principal Components Analyses ["factor analyses"] to see which SCs were most highly associated – The resulting 16 factors capture ~120 SCs. I also generated other factor solutions to gain perspective on the extent of association – e.g., 11, 21, and 28 factors.

13 Morillo, F., Bordons, M., and Gomez, I. (2003). Interdisciplinarity in Science: A Tentative Typology of Disciplines and Research Areas, *Journal of the American Society for Information Science and Technology*, v54 n13, p. 1237-1249.

- Another map of all 244 SCs based on cross-correlation with each other (a different way of looking at co-citation of SCs).
- An autocorrelation map. This is not as effectively clustered as the cross-correlation map.
- Two of the maps seem to work best: the cross-correlation map together with the 16-factor map. These shed light on the degree of association among the SCs that intuitively appear to belong together.
- For the 16 factors, I also generated the corresponding factor matrix and pasted the more inclusive top SCs (with loadings above ~0.2, noting natural gaps) from this into Excel. This extends the factors below the cutoff used in factor mapping.

Principles guiding domain formulation:
- Begin with an intuitively sensible grouping and examine consistency with the statistical clustering, using both cross-correlation and factor mapping.
- Seek relatively few domains, but each domain should generally share research norms within its SCs.
- Allow an "Other" category for SCs that really don't fit into the identified research domains well.

Assignment is not unambiguous; many SCs show some degree of relationship to different domains. But without providing further details, here is how FTA work locates in these resulting 12 larger domains.

Table 9 shows the aggregations. The last column is included to help gauge the scope of these research domains (note that the "engineering, mechanical & related" is quite discrete). The "# of FTA publications" shows where these 1018 papers are published. Because many journals are associated with multiple SCs, the "total" is considerably greater than 1018. The results are interesting:
- FTA is most strongly associated with the "*Quantitative Applications*" domain. [This includes SCs in Applied Math, Computer and Information Sciences, Electrical and Industrial Engineering, Operations Research, Management and Business, and so forth.]
- Yet, *FTA topics appear quite widely*, with over 100 in each of five additional research domains – in descending order: Social Sciences, Physical Sciences, Agricultural/Environmental Sciences, Earth Sciences & Civil Engineering, and Mechanical & Related Engineering.

Table 9 FTA Relationships with Research Domains

Research Domain	# of FTA Publications	# of Times FTA Papers Cite	# of SCs in the Domain
Quantitative Applications	624	6124	26
Social Sciences	275	1496	22
Physical Sciences	132	514	30
Life Sciences – Agricultural & Environmental	121	1165	26
Earth Sciences & Civil Engineering Related	112	711	17
Engineering – Mechanical & Related	105	290	4
Life Sciences – Biomedical	53	549	27
Behavioral Sciences	23	252	20
Other	19	76	18
Health Sciences	14	105	13
Medicine	3	130	23
Humanities	1	19	18
Total	1482	11431	244

The next column, "# of Times FTA Papers Cite," shows what research knowledge contributes prominently to FTA. The authors of the 1018 papers have referenced 11,431 prior journal articles that our journal-SC thesaurus was able to capture [so about 11 per paper]. Notable findings:

- Again, "*Quantitative Applications*" is the primary knowledge source referenced by FTA work
- As with the FTA publication pattern, Social Sciences is second.
- The richness of the reference base indicates that FTA does, indeed, integrate knowledge from a diverse range of sources. We find over 100 citations to papers from 10 of these 12 research domains.

Figure 4 shows the extent of diffusion of the major FTA topical themes (Table 6) across the Research Domains, based on publications (Table 9, "# of FTA publications"). [The research domains are scaled roughly proportionately to their frequency in this literature set.] This mapping suggests that FTA topics are not tightly bound to particular disciplines.

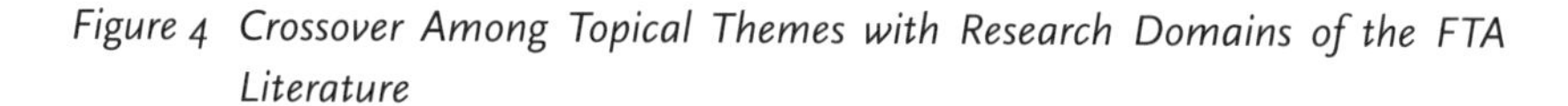

Figure 4 Crossover Among Topical Themes with Research Domains of the FTA Literature

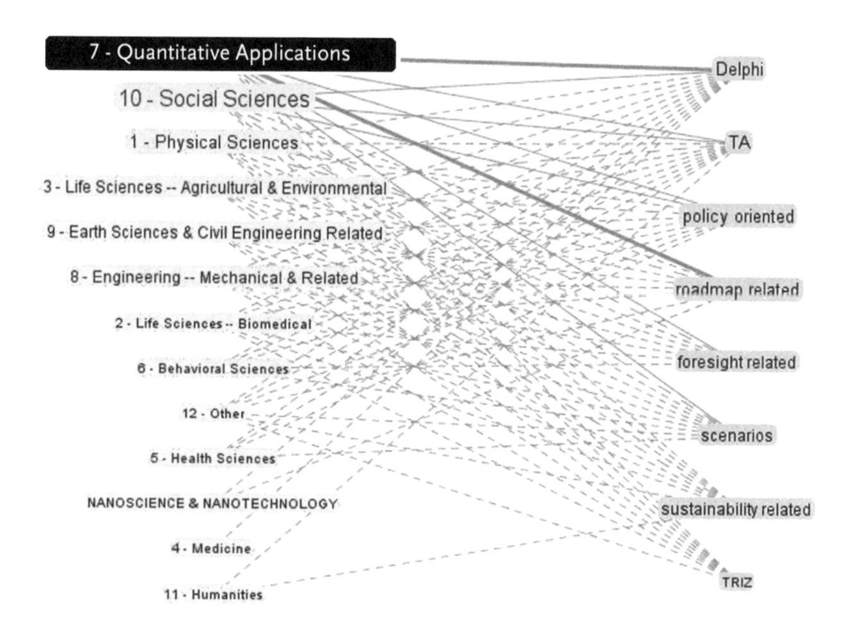

Table 10 parses these same data somewhat differently. This shows only those Research Domains with sizable numbers of FTA publications. It then normalizes the cell frequencies by the relative size of the respective Topic and Domain. Normalization reflects multiplying each cell's row value by its column value, and then summing for each of these cells. The values shown take the cell frequency (e.g., 115 for Delphi appearing in "Quantitative Applications") divided by the expected cell frequency (e.g., 177 papers mentioning Delphi times 624 Quantitative Applications papers). This is divided by the sum over all these cells, so that the percentages shown add to 1.
The cell values in bold show the Research Domains in which the given FTA Topic is relatively most used. For example, "sustainability related" themes are most prevalent in Agricultural/Environmental Life Sciences papers, whereas "TRIZ" appears proportionately most in Mechanical Engineering and Related SCs. So, while Figure 4 shows that topical themes are not tightly localized, Table 10 shows that, neither, are they evenly used across disciplines.

Table 10 Distribution of FTA Topical Themes across the Leading Research Domains

FTA Topics/Domains	Quantitative Applications	Social Sciences	Physical Sciences	Agri/Environ Science	Earth Sciences/Civil Engr rltd.	Mech. Engr & rltd.	Biomedical Sciences
Delphi	**1.71%**	**2.02%**	0.35%	**1.53%**	0.83%	0.79%	**2.27%**
TA	1.29%	2.15%	1.32%	2.64%	1.73%	2.49%	1.65%
policy oriented	1.50%	**2.78%**	0.47%	**2.95%**	**2.42%**	**2.23%**	1.16%
roadmap related	**1.94%**	0.89%	**4.40%**	0.43%	0.69%	1.35%	**1.95%**
foresight related	**1.94%**	**2.68%**	1.01%	0.85%	1.32%	**2.53%**	**1.95%**
scenarios	1.58%	**2.54%**	1.33%	**3.15%**	**1.99%**	1.21%	0.30%
sustainability related	0.82%	1.94%	1.20%	**6.69%**	2.82%	0.94%	1.12%
TRIZ	1.73%	0.31%	1.63%	0.36%	0.77%	**5.75%**	1.63%

6 Discussion

This paper has profiled the FTA research-oriented literature. It composed a search covering three Web of Science databases – Science Citation Index, Social Science Citation Index, and Arts & Humanities Citation Index. The resulting 1,018 articles published over a decade give a fair portrayal of that literature's emphases and trends.
Research-oriented FTA publication rates are modest, but the trend seems upward over the past few years (coincident with the initiation of the FTA seminar series). The U.S. leads in authorship, with a wide dispersion of author affiliations. Not surprisingly, university affiliations lead, but government and industry contribute strongly to this literature. Journals concentrate in technology management, led by *Technological Forecasting & Social Change*. Eight prominent topical themes are identified and plotted over time – Delphi and sustainability analyses show recent surges in attention.
The special interest is to see how the FTA literature distributes across disciplines. To do that, I build upon the Subject Categories (SCs) defined by Web of Science. The leading outlets for FTA articles reside in the Planning & Development, Business, and Management SCs. Those three, along with

Economics and Operations Research & Management Science, are most cited by the FTA articles. "Research domains" built by clustering the SCs into related groupings help show a larger picture. Publication of the FTA literature centers in the "Quantitative Applications" domain, with a second concentration in "Social Sciences", and significant participation in a range of Life Sciences, Physical Sciences, and Engineering domains. Equally interesting, the literature FTA papers reference show a similar pattern, drawing upon diverse sources led by Quantitative Applications, with Social Sciences second. Table 10 shows the prevalence of FTA topical themes across the leading research domains for FTA publication.

I hope this provides useful empirical background for the further examination of how different disciplines approach FTA.

Acknowledgements

I would particularly like to thank my National Academies Keck Futures Initiative colleagues – Anne Heberger and David Roessner – and my Search Technology colleagues – David J. Schoeneck and Webb Myers – for their contributions in developing the ISI Subject Category and Domain analyses. Henry Small of Thomson Scientific (ISI) kindly provided the underlying journal to subject category thesauri.

12 Futures research and science: summary and some afterthoughts

Patrick van der Duin

If there is one science, scientific discipline, practice or even art that has a future, then it must be futures research. Not only because there is a lot of future ahead of us, but also because the political, economical, social and above all, environmental problems that the world currently faces demand a different, some would even argue, a completely different approach to the future. Some might argue that a different *approach* is not enough, because what the world needs is action – right now – rather than merely thinking and theorizing about the future. However, theory and action cannot be separated, especially not in futures research, which can be considered an applied science. The urgent need to change our way of living, as suggested by an increasing number of citizens and even politicians should, therefore, be grounded in a different way of thinking. I am not suggesting a specific direction, since that is not my area of expertise (and I would be hard pressed to choose a direction) and also because that is not the main topic of this book, which advocates a much more future-oriented way of thinking as well as acting. The good news is that there seems to be an increasing interest in issues to do with the future, and some futures researchers (especially the trend-watchers and futurologists) may experience this in an increase in their 'business'. However, to satisfy the increasing need for looking to the future and to make sure that futures researchers can live up to the heightened expectations facing them as essential guideposts to a better future (whatever that may mean), they need to improve the quality and with it the relevance of their work. One way of doing this is by reconsidering the (historical) link between futures research and other scientific disciplines, because it provides an important context in which futures researchers operate, regardless of the fact that many questions about the future are not even answered by futures researchers themselves, but are posed directly to

specialists in other scientific disciplines. Thus, an awareness of the role the concept of the future plays in others sciences, from which much knowledge and information is being extracted, will help futures researchers to do their work. However, this book not only has an instrumental goal: the question of how the future is being addressed in different sciences is an interesting one in itself. And I am glad that the authors of this book share my opinion in that respect and have addressed this issue thoroughly.
Although each chapter of this book speaks for itself, I will make an attempt to summarize its main points by repeating those sentences from each chapter that I regard as central to that chapter, adding my own interpretation and providing further elaboration. In doing so, I attempt to address the topic of this book in the best way possible.

In Chapter 2 Peter Hayward looked at the topic of this book from the science of psychology. As far as I am concerned, the central remark in his contribution was that "(t)he future is not a 'received view' but is, instead a 'constructed view'". I think that this means that the future is not presented to us on a platter, but that we interact closely with the future. Thus, the future is not the same for everyone, and it depends on the personal interpretation of each of us. Each of us looks at the parts of 'the' future that is interesting or relevant to us, interpreting the future and synchronising our actions accordingly. There is no such thing as 'the' future because there is no such thing as 'the' man or woman. Of course there are limits to the individuality and uniqueness of mankind, but looking to the future is often a personal affair. It is not for nothing that the famous Delphi method starts by interviewing experts individually. It would appear that researchers want to know first what individuals think about the future and ways to realise it, before confronting them with the opinions of others and turning 'the' future into the collective outcome of the various opinions. In organisations problems may also arise when various people have their own views with regard to the future, which may make mutual cooperation. In terms of the coordination of business activities it is important to have a certain consensus about what the organisation wants to be in the future and how that future can be secured. At the same time it is important not to sacrifice every (differing) vision of the future to 'group think'. Finding the right balance between a coherent vision of the future and divergent visions ensures a sufficient consistency between the business activities and having an 'open' approach to the future.

In Chapter 3, Eleonora Barbieri Masini looked at the future from a human and social perspective. I consider the following sentence to be central to her contribution: "...Sociological concepts, theories and methods have enriched and fertilised the futures thinkers toolbox and their conceptual repertoire". Where Hayward in Chapter 2 addresses the psychological (individual) aspects of looking to the future, Masini emphasizes the social dimension of the future. The quote also shows that future studies do not (only) focus on tools with which to try and predict the future, and that there is more than merely an instrumental approach. The future is not the outcome of the application of a certain method by a select group of people. It is a broader activity within which the process of thinking about the future assumes a central importance. To this, we can add the future is not a neutral entity that can be explained or predicted via scientific tools. Although personally I consider the application of methods of futures research as a part of looking to the future, Masini has a valid point when she argues that every vision of the future has a normative and ethical component. Later in the chapter, she argues that "...It is hence important to repeat that FS are not only a discipline but also a moral duty". I want to add to this that the 'scientific' side of futures research (predicting and exploring via methods and tools) and normative (human and ethical) side of van future studies can be mutually beneficial. After all, by applying methods of futures research we can decide whether or not we are moving towards a future that is ours. Futures research methods can also be used to determine whether a formulated future can be realised. However, let me be clear as to the order: futures research that is inspired by ethical, social and human values precedes the application of (scientific) methods of futures research. This makes it clear that the (predicted or explored) future is not something that happens to us. It is something of which we are a part and for which we are all responsible.

In Chapter 4, Cornelius Hazeu established a link between economy and the future. That link is not unproblematic. According to Hazeu, economists have a problem considering the longer term: "The most difficult thing, including for economists, is the real long term". Apparently, economists become nervous when their statements regarding the future or their models are insufficiently statistically. And because economists have to maintain a tradition in conducting quantitative and mathematical research, in which they often like to compare themselves to physics, they are weary of the longer term and limit themselves to the short term. Perhaps economists prefer being 'precisely wrong' than 'roughly right', although that statement is not completely fair, in light of the accuracy of many short-term predictions and recommendations. Thus, focussing on the short term and even on the present does not have to be wrong in itself. According to Hazeu, economists realise full well that making the right decisions in the present is important when we approach the future: "The *future* might be by definition unknowable, but this does not take away the constant need of governments, businesses, citizens/consumers to seek guidelines with which to underpin the decisions that they have to take *today*". Despite these correct words, this does put economists in a dilemma. It is precisely in order to make the right decisions in the present that we need knowledge about the future. Economists seem to agree about the time horizon of that future: as short as possible, or so it would seem. However, many economic decisions involve a longer time frame that falls outside the margins of error of economic models. But the long term cannot be seen as a sum of the short term. Incremental thinking makes us shuffle step by step towards the abyss. It is not for nothing that economists are very interested in research into long-term economic cycles (or waves) that apply less to the short term but that do provide insight into developments in the longer term. It allows economists to connect the short term to the longer term in a meaningful way. This is not to say that economists have a clear enough view of the future (in the short and longer term), because the economy is more than a set of waves or an extensive set of data. In essence, economy has to do with the behaviour of individuals and organisations, which means that the economic system is open enough to keep the future open.

In Chapter 5, Joseph Voros addressed the philosophical foundations van futures research. As far as Voros is concerned, every future research should pay attention to philosophical aspects: "...the philosophical bases of enquiry – the foundational assumptions and fundamental presuppositions about reality, knowledge and method, to name but a few – must be explicitly taken into account in order to ensure that the form and approach of an enquiry is appropriate to the purpose and domain of the enquiry". What touches me most personally is the word 'explicitly', which allows the 'real' futures researchers to distinguish themselves from the fake or bad futures researchers. With these latter two qualifications I am not referring to the 'man in the street' who looks to the future because he has to make decisions that concern the future (for instance education, marriage, buying a house, etc.) but rather to the gurus, trend watchers, and many futurologists who often make a convincing impression, but whose statements are often not based on verifiable assumptions and analyses. It is precisely by being open and providing others insight into the process of futures research that the futures researcher can eliminate his or her own mysticism. At first sight, this may appear to be disadvantage, but in the end it will increase the appreciation for the futures researcher and his or her work. Not only because of his personal openness, transparency and honesty, but also because it now becomes clear how a futures researcher has arrived at his or her results. Perhaps many people will continue to believe that futures research gaze into crystal balls, but at least everyone is invited to take a look for themselves. Rather than being removed from his ivory tower as a result, I believe the futures researchers will be placed upon a pedestal. Insight and access often generate appreciation and admiration. Perhaps not everybody will be interested in the philosophical foundations of futures research, but it will surely reinforce, not only the knowledge base of futures research, but also the position of the futures researcher.

In Chapter 6, Sohail Inayatullah described what the meaning can be of macrohistory and macrohistorians for the future. In my view the opening sentence of his contribution is also its core: “Through its delineation of the patterns of history, macrohistory gives a structure to the often fanciful visions of futurists. Macrohistory gives us the weight of history balancing the pull of the image of the future. Yet like future studies, it seeks to transform past, present and future not merely reflect upon social space and time”. Macrohistory provides the futures researcher or futurologist with the opportunity to place his or her own thoughts about the future in a historical framework, As a result, predictions or explorations no longer exist in isolation (linked at best to the present and a little bit of the past), but instead become part of a historical pattern that, in order to be truly worthy of that epithet, need to have a certain predictive value. A historical pattern that stops in the present and does not continue into the future is of little value to the futures researcher. An important lesson that the futures researcher can learn from the macrohistorian is to look for change, and to view matters in a larger perspective. A historical pattern suggests change and dynamism, and they are also important ingredients of futures research. Unfortunately, futures studies all too often describe a (steady) end-state, at best with a description of the path leading to that end-state. In particular utopias excel in describing a future that is static, with little hope for change, much less improvement. There is no progress anyway, and especially in the case of dystopias this may be seen as a problem. In a pleasant utopia there is little need for change and even dullness is embraced as preferable to uncertainty. By thinking in patterns, in *changing* systems, the futures researcher, in a manner of speaking, can look beyond his futures study. This does not mean that the time horizon of a futures study is an empty concept, but like the presence contains *seeds of change* that lead to a future that is different from the presence, the future itself has certain (future) *seeds of change* that will lead to a different, new future. From personal experience I know that many scenario studies (like many utopias) neglect to describe a certain tension. The future never stops, is in principle infinite, and as a result it is interesting to discover possible future tensions, dilemma's, issues, opportunities, and problems. Perhaps this may help resolve the problem that many futures studies tell us more about the present than about the future. In fact, it means the futures researcher is developing a kind of *meta-future*. In other words, not only about the future or futures as such, but also about the future that will be predicted or explored in the future. I admit that it is hard enough investigating the future in the presence that exploring *meta-futures* is virtually impossible. Nevertheless, I think it would improve the quality of futures research if we were not to consider the future as a static and finite state, but instead decide to look also at future *seeds of change*.

In Chapter 7, Ela Krawczyk discussed the relationship between geography and looking to the future. As far as geographers are concerned, the future is a clear addition to their field of expertise: "Many geographers see time as the fourth dimension of space (...); however, their perception of the relationship between space and time varies". Krawczyk provides a clear overview of the history of the role of the future in geography. The historical line she suggests runs in parallel to that of futures research itself, which suggests a close relationship between futures research and geography. Even her recommendation with regard to the application of futures research in geography will please many futures researchers: "Although, clearly, futures methods have been increasingly incorporated into planners' toolkit, I would argue that there is also a strong need to develop a 'futures mindset' amongst planners and decision makers that would change their underlying perceptions about the future and consequently help to use these methods to their full potential". Geography and future are connected by the concept of 'planning'. This gives looking to the future within geography a clear instrumental function. As far as geographers are concerned, looking to the future is not a goal in itself, but instead it provides important input to geography in general and planning in particular. The direction of the relationship between geography and futures research does not immediately make it clear whether geography has influenced the way of looking to the future or whether the development of futures research is autonomous, in other words, independent of the developments in geography. I would not presume to argue that futures research has had a (decisive) influence on the developments in geography and I suspect that Krawczyk agrees with me. The question what the exact direction of the influence is, is not really all that interesting anyway. What is more important is that geographers pay attention to time in general and the future in particular, and that this case shows us a good example of symbiosis between two scientific disciplines.

In Chapter 8, Graham May wrote about the relationship between futures research and the natural environment. Although nature cannot be considered a scientific discipline[1], it is a crucial subject for futures research. In fact, it is of vital importance to mankind that we pay sufficient attention to this subject. As May puts it: "For the survival of the human race there is no more important issue than the environment that supports life on earth". This places a great burden on the shoulders of futures researchers, although our survival concerns us all. Futures researchers can certainly play a crucial role in developing a sense of urgency. By presenting us with a future that would be fatal to us, or by pointing at the detrimental consequences in the present, futures researchers can fulfil the role of warning messengers. It is important that they not assume the role of scaremongers, because what is worse than a doomsday scenario that comes true is a doomsday scenario that nobody believes. It is precisely with this subject that the multidisciplinary and diversity of futures research emerges. Nature touches all aspects of life, natural and social, and it may very well be the outcome of actions within all those domains. In May's words: "The complexity involved in thinking about the environmental future is further emphasised by the links that each of these issues gave for the economic, technological, social and political developments". This does not make the futures researcher's work any easier, but it does make it more important. It is especially the futures researcher whose work is interdisciplinary, multidisciplinary en transdisciplinary (see Chapter 1) who can play a central role. No longer as the bearer of bad tidings, but (with it) also as an instigator of ideas about change and future alternatives. A good futures researcher does not monopolise this subject, but instead brings out the best in representatives of other scientific disciplines, summarises what they have to say and presents the right answers to the question of 'what-if'. In this respect, the futures researcher acts primarily as a future *process* expert rather than a future *content* expert (see Chapter 1).

In Chapter 9, Patrick van der Duin en Erik den Hartigh connected management science to the future. Like Ela Krawczyk (Chapter 7 on geography and the future),

1 Unlike biology, of course, but that is not the subject of that chapter.

they see futures research as a servant to other science: "For the management sciences, outcomes of futures research are not goals in themselves but provide inputs to decision-making processes within companies". Although futures researchers are presented here as servants to another science, that does not mean that futures researchers serve the decision-makers who often are the same people that pay the futures researchers. This relationship between futures researcher and decision-maker jeopardises the independence of the futures researcher, which in practice usually means that the futures researcher has to deliver a result that is convenient to the decision-maker, and one toward which the decision-maker actively urges the researcher in the course of the research process. This means that the futures researcher on the one hand will feel the pressure of the decision-makers, who will subtly remind the researcher who it is that pays the bills, while on the other hand the futures researcher will want to tell and write what he or see perceives to be right. A 'good' decision-maker will insist that 'his' futures researcher give his or her opinion, but alas this type of decision-maker is not the prevailing type. Perhaps this dilemma will not play a major role for other futures researchers, because they say one should always set one's own course. And although this is a very valid stance, we have to admit that many futures researchers do in fact face these kinds of problems. The bad news, therefore, is that futures researchers may become stuck, but the good news is that the decision-makers in question at least have an eye for the future. In many cases, organisations, in particular those of a commercial ilk, find it hard enough as it is to find time to look to the future at all. Van der Duin and Den Hartigh: "...the increasing influence of shareholders, requiring companies to provide quarterly report of their earnings, often results in a more short-term view of the future". If we look at it that way, futures researchers battle on two consecutive fronts. On the first front, they need to fight for the organisation's attention for the future, and secondly they have to make sure that they themselves leave their mark on the futures study. I would be hard pressed to say which of these battles is more important.

In Chapter 10, Dap Hartmann described how astronomers deal with the future. His opening line does not bode well: "Astronomers only observe the past". But because of the immutable lows of physics, based on historical observations, it is possible to produce beautiful and correct predictions. Hartmann's is the only contribution that establishes a link between futures research and a natural science.[2] I can imagine that many futures researchers, and I regard futures

2 Some would perhaps argue that economy is a natural science, or at the least a quantitative science, but I am not one of them.

research as a(n) (applied) social science, look with a fair amount of jealousy at the natural sciences in general and astronomy in particular. Although predicting the future is certainly not the only goal of futures research, it would be worth a lot if social phenomena could be predicted with the same level of accuracy as in the natural sciences. The two main differences between natural and social sciences are that social reality is an open system and natural reality a closed one, and that the interaction between the actor and the system is far greater in social reality than it is in the natural one. As far as the latter difference is concerned, it becomes clear that the interaction between a prediction or exploration and the action and reaction on the part of the actors (citizens, organisations) becomes important. Whereas the predicted or explored natural reality is detached from what we do (or do not do), in the case of the social reality the future situation not only depends on a prediction or exploration, but also on the behaviour of the actors. Not only the future situation, but the prediction or exploration itself determines behaviour that in turn affects the future situation, resulting in self-fulfilling or self-denying prophecies, the social science's equivalent of the Heisenberg principle. Although this is a factor that makes the futures researcher's job both difficult and interesting, it does not remove the jealousy with which he or she feels towards astronomers whose predictions will not effect celestial bodies.

In Chapter 11, Alan Porter presented an overview of futures research (or FTA) literature with regards to, among other things, the number of publications, the magazines and the organisations that are active in this area. His conclusion is: "Research-oriented FTA publication rates are modest, but the trend seems upward over the past few years (coincident with the initiation of the FTA seminar series). The U.S. leads in authorship, with a wide dispersion of author affiliations". Despite this good news, the question remains why it is that we see this trend. I think it can be partly explained by the fact that technological developments appear to have an acceleration, and that technology is once again more often seen as a benefactor. I think we also need to put the good news into perspective, because publishing about the future is not the same as thinking about and acting on the future. This raises the question how we can measure the effect and impact of futures research. Porter shows us how to measure and analyse the number of publications about the future, but determining the impact of futures research on decisions is another, more tricky, matter. Perhaps we need to look at the input as well as the output to determine the real role and influence of futures research. Data concerning the number of publications, studies or futures researchers represent one side of the coin. We also need to gain insight into the way certain decisions or their implementations can be related to a futures study. We can, for instance, look at the extent to which futures research manages to put a certain subject on the agenda. The problem with the implementation of futures research is that it often happens implicitly. Futures studies enter the minds of people (decision-makers, policy-makers), and although this is in itself a good thing, it makes it harder to determine the relationship between the futures study and the eventual decision. This explicitness is one of the main problems in the implementation of futures researchers, apart from the fact that in many cases other (non-future-related) aspects also play a role, which makes it harder to establish the difference between organisations that engage in futures research and organisations that do not. These two problems are more difficult to resolve that the classic problem that one needs to wait for the future to arrive before being able to determine the usefulness of futures research. Nevertheless, in my view, these two problems are among the most crucial problems facing futures research, and they will play an important role in the future development of futures research.

At the end of the book, the question is whether it is possible to draw a general conclusion. This is a difficult question, because it is the purpose of this book to show the diversity of futures research by connecting it to various scientific disciplines. Several authors have placed the subject in a historical context, showing us that futures research is older and less modern than one might assume. Of course there have always been prophets and soothsayers, but in the various sciences future-related aspects have been discussed. Other authors have emphasized the instrumental function of futures research, presenting futures research as a kind of auxiliary force that may support a scientific discipline.

On the basis of these two findings we can say that futures research has a long and versatile history and that it has evolved into an independent scientific discipline that may be of service to other scientific disciplines. Many futures researchers ask themselves the meta-question what that future of their discipline is. Although this book does not address that question, we can say that this kind of reflection may be crucially important to the future development, quality and position of futures research. I hope this book has provided a contribution in that.

About the authors

Eleonora Barbieri Masini was borne in Guatemala. She has a Degree in Law and Sociology and a PhD in Social Change. She has taught Futures Studies at the Faculty of Social Sciences, Gregorian University in Rome from 1976 to 2004 and Human Ecology from 1994 to 2004. She has been tutoring doctoral students in Futures Studies from 1995. From 2004 she is professor Emeritus at the Faculty of Social Sciences, Gregorian University. She also has been Secretary General for five years and President for ten years of the World Futures Studies Federation and is member of the Club of Rome since 1975 and from 2004 Honorary member. Next to this all she is member of the Board of Directors of the World Future Society. In 1998 she became Doctor Honoris Causa at the University of Economic Sciences of Budapest. Eleonora is the author of books and publications on futures research among which "Visions of Desirable Societies" (1983), "Women, Household and Age" (1991), "Why Futures Studies?" (1993), "La prevision umana y social" (1993), "Penser le futur" (2000) as well as various recent articles for *Futures*. Eleonora can be contacted at: e-mail e.masini@mclink.it

Erik den Hartigh is an assistant professor at the Department of Technology, Strategy and Entrepreneurship at the faculty of Technology, Policy and Management of the Delft University of Technology. He teaches in corporate strategy and innovation systems. His main research area is company strategy for dealing with increasing returns: positive feedback effects in firms and markets. He is also a consultant with TVA Developments, a small consultancy company for strategic transformation.

Dap Hartmann (1960) got his Ph.D. in Astronomy (on the distribution of atomic hydrogen in the Milky Way Galaxy) from the University of Leiden (1994). He was a visiting scientist in observational astronomy at the Harvard-Smithsonian Center for Astrophysics, the University of Bonn, and the Max

Planck Institut für Radioastronomie. He is the author of the *Atlas of Galactic Neutral Hydrogen*, published by Cambridge University Press.
Since 2003, he is assistant professor at Delft University of Technology in the faculty of Technology, Policy and Management (section Technology, Strategy & Entrepreneurship). His current field of research is innovation management and entrepreneurship. He is co-author of *The Cyclic Nature of Innovation. Connecting Hard Sciences with Soft Values*, published by Elsevier. Dap Hartmann writes about science and technology for *NRC Handelsblad* and various other periodicals, and has made four radio documentaries on science and technology for Dutch national radio. He is a fellow of the Center for Inquiry for the Low Countries.

Peter Hayward is a foresight practitioner and academic. He specialises in helping organisations and communities act creatively with the future in mind – learning to see the world differently and find hopeful and inspiring futures. Peter writes, speaks and leads workshops. His approach is based in systemic understandings of the present and the creation of powerful images of the future. He is the Program Director of the Masters of Strategic Foresight that is offered at Swinburne University in Melbourne. That is the only post-graduate qualification in foresight that is offered in Australia. He began his career as an accountant and economist working for the Australian Taxation Office. Increasingly he became interested in how change happens and then how people create change. After studying Systems Thinking at the University of Western Sydney he became interested in the idea of how individuals and organisations can create positive futures. Peter has written on the topics of psychology, systems thinking and foresight. He has published The Moral Impediments to Foresight Action, Facilitating Foresight, Foresight in Everyday Life and Foresight as a Catalyst for Change. His PhD was From Individual to Social Foresight and studied how foresight develops in individuals. Peter is also an exponent of Integral Theory, as espoused by Ken Wilber, and he now employs that framework to deepen organisations understandings of their future options.

Cornelius Hazeu, PhD, has his own Office for Institutional-Economic Research and Advice. Before he has been a staff member of the Scientific Council for Government Policy in the Netherlands and has been an associate professor in general economics at the Erasmus University of Rotterdam. His fields of interest and expertise are: (1) institutional economics, (2) future research, (3) higher education and science, and (4) public finance, social security and the welfare state. For contact: hazeu@wanadoo.nl

Sohail Inayatullah is a political scientist/futurist associated with Tamkang University, Taiwan (Graduate Institute for Futures Studies), University of the Sunshine Coast (Faculty of Arts and Social Sciences), and Prout College (www.proutcollege.org). He has authored/edited eighteen books and cdroms including; Neohumanistic Educational Futures; Questioning the Future; The Causal Layered Analysis Reader, and The University in Transformation. Inayatullah has authored over 300 referred journal articles, book chapters and magazine editorials. He is a theme editor of the UNESCO Encyclopedia of Life Support Systems and has contributed articles to the Macmillan Encyclopedia of Futures Studies and the Routledge Encyclopedia of Indian Philosophy

Ela Krawczyk, PhD, MA. Since July 2006, Ela has been working as a Postdoctoral Fellow at The Futures Academy, Dublin Institute of Technology, where she has been developing a 'methodological knowledge base' for the application of futures methods in urban planning processes. Her expertise lies in using futures approaches in the wider built environment. She is also very interested in the processes shaping urban places and issues related to sustainability. Ela's doctoral thesis explored futures thinking in the urban planning context, using Dublin as a case study. Ela also has a Masters Degree in Social-Economic Geography from the Jagiellonian University, Krakow, Poland. From August 2005 to July 2007 she held a position of Assistant Secretary-General of the World Futures Studies Federation.
The Futures Academy, DIT, Bolton St., Dublin 1, Ireland. T: +353-1-4023749, E: elzbieta.krawczyk@dit.ie

Graham May BA, MA, FRSA formed Futures Skills: Education and Training in Foresight and began working with the Foresight Group in PREST at Manchester University in January 2003. Prior to that he was Principal Lecturer in Futures Research at Leeds Metropolitan University. Graham is author of *"The Future is Ours: Foreseeing, Managing and Creating the Future"* Adamantine Press and Praeger Twenty-First Century Studies, 1996, and of papers and articles in a range of publications including *Futures, Futures Research Quarterly (USA), the Journal of Futures Studies (Taiwan), the American Behavioral Scientist, and Seminar (India)* and chapters in *"Tomorrow's Company"*, *"Sustainability and the Future"*, *"The Education Yearbook 1998"*, *"Rescuing all our futures"* and *"Advancing Futures"*. From 1996 until 2002 he was Course Leader of the Masters degree in Foresight and Futures Studies, and has extensive experience teaching Futures at undergraduate and post-graduate levels. He has been advisor to a number of companies and local, regional and national government, contributed to British

Council Workshops in Moscow and Kiev and edited a distance learning course in Technology Foresight for the United Nations Industrial Development Organisation UNIDO. He played a leading role with the Local Government Association in producing the *FuturesToolkit*. At PREST he has been involved in a number of projects funded by European organisations including sectoral studies for the European Foundation for Living and Working Conditions and the annual Foresight Course.
He is a Fellow of the Royal Society of Arts, a member of the Advisory Council and Professional Member of the World Future Society, a member of the of the World Futures Studies Federation, member of the Editorial Board of *Foresight, and* the *Journal of Futures Studies,* a member of the Tomorrow Network and Council Member of the UK Futurists' Network. He was formerly Convenor of the UK Futures Group and a member of the Construction Panel of the UK Government's Foresight Programme.

Alan Porter is Director of R&D for Search Technology, Inc., Norcross, GA. He is also Professor Emeritus of Industrial & Systems Engineering, and of Public Policy at Georgia Tech, where he remains with the Technology Policy and Assessment Center. He is author of some 200 articles and books, including *Tech Mining* (Wiley, 2005). He is pursuing ways to exploit science & technology information to generate intelligence on emerging technologies. He has worked with various organizations to improve their technology watch capabilities. Recent work has keyed on ways to generate Quick Technical Intelligence Products (QTIP) to support organizational decision processes. He received a B.S. in Chemical Engineering from Caltech, 1967; and a PhD in Engineering Psychology from UCLA, 1972.

Patrick van der Duin (1970) was trained as a macro-economist at the University of Amsterdam. He worked for six years as a researcher at KPN Research, where he conducted several studies on the future of telecommunication. In 2006, he did his PhD at Delft University of Technology on the relationship between futures research and innovation. His fields of interest are (qualitative) futures research in commercial organisations, evaluating studies of the future, and innovation management. Patrick works as an assistant professor at Delft University of Technology. He has published articles in Foresight, Futures, Technological Forecasting & Social Change and the International Journal of Technology Management.

Joseph Voros began his career as a physicist – he holds a PhD in theoretical physics, during which he worked on mathematical extensions to the General Theory of Relativity – followed by several years in Internet-related companies, before becoming a professional futurist. His professional interests are broadly multi-disciplinary, and his main research interests are similarly varied. Currently, these include: 'integral' approaches to futures research; the 'epistemological problem' in futures research; theoretical bases of futures methodologies; 'strategic foresight' in organisations and society; the interplay and overlap between futures thinking, creativity, strategy and policy; and the long-term future of humankind.

About the WRR

How will European citizens make a living in twenty years from today? What will the labour market and the welfare state look like in an age of rapid demographic ageing? How can we stimulate citizens, companies and private and public institutions to take on greater responsibility for issues of physical safety in their immediate environment? What does political Islam have in store for the Netherlands, Europe and the world at large? How can long-term investments in key infrastructures be safeguarded under conditions of market liberalisation? What place does Europe have in tomorrow's world and what status will the Netherlands and other member states occupy in an enlarged European Union? These are all key questions for futures studies. Our task at the Netherlands' Scientific Council for Government Policy (WRR) is to render advice to the Dutch government about these future questions of great public interest on the basis of policy oriented academic research.

What the WRR tries to stimulate above all is an in depth debate over preconceived policy assumptions in order to create a window of opportunity for placing alternative policy orientations on the political agenda. Over the past 35 years we have however learnt a few evocative and sobering lessons for futures studies. First of all, the future always transcends existing policy sectors and silos. As a consequence, futures analysis needs to be approached from different disciplinary angles. The most important academic prerequisite for futures researchers is not so much that they need to be interdisciplinary. More so: they have to be able to transcend their original academic upbringing, to develop an eager curiosity for what goes on in other academic and scientific quarters. For us at the WRR, a passion for public policy and a form of ideational imagination, are a sine qua non for futures studies, next to sound research competencies.

In 1977, the WRR took on a large-scale, all embracing scenario study of what the future had in store for Dutch society 25 years hence. A recent evaluation of the 1977 report The Next Twenty-five Years, exploring the future brought to the fore some sobering conclusions. The 2004 study 'Twenty-five years later, the survey of the future 1977 as a learning process', revealed that some predictions were true, but that a lot of them were not. Our projections on the state of the economy were fairly correct, but we did not at all fathom the collapse communism and its consequences for the European Union and the geopolitical context we are experiencing today. In futures research by and large, there is a tendency to think in terms of path-breaking policy at moments of structural disequilibria. Path-breaking change takes place at moments of institutional rupture during which pre-existing institutions are reformed in both their design and the 'hierarchy of goals'. To be sure, one can expect that crisis conditions open up opportunities for policy redesign, but crisis leading to convulsive historical ruptures are extremely hard to imagine, let alone predict. Moreover, we know better than ever before that key features of public policy and formal institutions are pretty change resistant. Strong political actors which have invested much of their in existing institutional repertoires often militate strongly against significant policy change. Perhaps the most interesting avenue of futures research is conjecturing institutional change that is both incremental and transformative. Changes occurring at the margins of policy can cumulative and over time fundamentally alter existing policy repertoires, with major departures from original policy functions. Take the example of the national welfare state in Europe. In the 1950s the expansion of social insurance coverage was targeted to productive male breadwinner families. The 1960s and 1970s prompted the universalisation of social security, with the introduction of social assistance. In the 1990s new activation programs and the expansion of social services, family and women friendly leave schemes, vocational training and education, informed by the new policy logic of social investments, transformed to the contours of the welfare state were initiated. In the process the original design of the industrial welfare state was transformed to a new post-industrial edifice informed by the new policy logic of social investment. This sort of redefinition of social policy purpose is an interesting example of self-transformative incremental institutional change. Self-transformative incremental policy change occurs in the absence of, rather than because of, major institutional collapse or breakdown. It is however important to bear in mind that only the long shadow of history provides the adequate backdrop against which to better observe and assess these kinds of evolutionary transformations.

Prof. dr. W.B.H.J. van de Donk, Chairman of the WRR

About Alares

Dutch society has gone through major changes in recent years. Some changes – societal disturbance and the shake-up of the political environment – were not predicted and had therefore an unexpected and negative impact. There is no reason to assume that such changes were one-off; it seems that since 9/11 'the unpredictable is now predictable'. Other developments, such as the noticeable impact of an ageing population, the increasing pressure on the labour market and the discussions about the pension-age are less unexpected, but do have nevertheless far reaching consequences as well. Issues concerning education and mobility have been subject of fierce debate for a longer time now, without any sight of solutions on the short term. All these developments have an (increasing) influence on all parts of society; citizens, businesses and authorities. Dutch citizens for example have another role nowadays. Partly because he/she actively requires it, partly because the governmental authorities more and more withdraw from their traditional public services. This leads to the situation that in the fields of social security and healthcare every citizen is required act independently and to show more initiative than some years ago.

It's necessary that all parties involved get a better understanding of these developments. And it's also important to accept that this process of understanding requires a new approach. The new approach includes the awareness that trans-sectored innovation is essential. The role and influence of innovation for a knowledge driven economy is widely recognized today. Trans-sectored thinking however is less well known: it is rarely used within discussions about innovation. But as the problems in one specific sector are often comparable to those in other sectors it's necessary that a sector-crossing approach of innovation is stimulated.

Alares, an innovative advisory company, that works for government, social sectors and businesses, endorses that it's important to be aware of actual themes in today's complex society. We invest in and encourage research concerning the impact of future developments on all aspects of society. Furthermore we participate actively in various social discussions about innovation and trans-sectored knowledge sharing. This external orientation, combined with the creativity and commercial capacities of its advisors, makes us an organization that holds an unique position in fields like future management, innovation strategy and program management. We therefore happily support this book on futures research and will use its results in our daily working field. To make Dutch society more future proof!

R.J.M. van Oirschot MBA, director of Alares